WITHDRAWN

CAMBRIDGE EARTH SCIENCE SERIES

*Editors:* A. H. Cook, W. B. Harland, N. F. Hughes, A. Putnis, J. C. Sclater and M. R. A. Thomson

The dawn of animal life
*A biohistorical study*

Reconstruction of some components of the Ediacara fauna. The large rooted leaves in the centre and left represent *Charniodiscus arboreus*, the pair on the right are other Charniidae. Above are various medusae: *Rugoconites, Ediacaria, Mawsonites*; a large *Brachina* is to the right of *Charniodiscus*. *Dickinsonia elongata* is shown swimming in front of it, other species are on the sea floor, as are *Tribrachidium* and *Conomedusites* (lower margin) and *Cyclomedusa* (cut by right margin). *Spriggina* (below) and *Parvancorina* (middle) are shown swimming near the margins of *Charniodiscus*. Drawing made by Robert Allen from sketches and instructions by Dr Mary Wade, Queensland Museum, Brisbane. Publication authorized by Dr Alan Bartholomai, Director.

# The dawn of animal life
## A biohistorical study

MARTIN F. GLAESSNER
*University of Adelaide, Adelaide, South Australia*

In practice, speculations about the past,
if they are not to be entirely idle,
must relate to the traces which
the past has left.
A. J. Ayer,
*The central questions of philosophy*

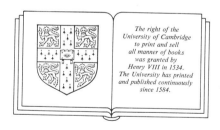

CAMBRIDGE UNIVERSITY PRESS
*Cambridge*
*London   New York   New Rochelle*
*Melbourne   Sydney*

Published by the Press Syndicate of the University of Cambridge
The Pitt Building, Trumpington Street, Cambridge CB2 1RP
32 East 57th Street, New York, NY 10022, USA
296 Beaconsfield Parade, Middle Park, Melbourne 3206, Australia

© Cambridge University Press 1984

First published 1984

Printed in Great Britain by the University Press, Cambridge

Library of Congress catalogue card number: 83-5188

*British Library Cataloguing in Publication Data*
Glaessner, Martin F.
The dawn of animal life. – (Cambridge earth science series)
1. Paleontology 2. Evolution
I. Title
560    QE711.2

ISBN 0521 23507 3

# Contents

*List of figures and tables*　vii

*Preface*　ix

**1　Precambrian life and its environment: a review**　1
1.1　Theories about the origin of life and its environment　1
1.2　The Precambrian, its subdivisions, and the dating of biohistoric events　5
1.3　The preservation and recognition of organic remains in rocks: chemofossils, fossil organisms, pseudofossils　9
1.4　The cell: organization and life processes. The 'kingdoms' of organisms　13
1.5　Proterozoic fossils and environments　22
1.6　Origins and early differentiation of the Metazoa　30

**2　The Ediacarian faunal assemblages: discovery, composition, significance**　43
2.1　Discoveries at Ediacara　43
2.2　Palaeoenvironment and fossilization　47
2.3　Composition of the fauna　50
2.4　Palaeobiology　65
2.5　Other occurrences of Late Precambrian Metazoa in Australia　69
2.6　The Nama fauna of southwestern Africa (Namibia) and possible equivalents in South America　73
2.7　Late Precambrian Metazoa from the Northern Hemisphere　83
2.8　The concept of the Ediacarian fauna　100

**3　The Precambrian diversification of the Metazoa in the light of palaeozoology**　106
3.1　The significance of the Ediacarian fauna for metazoan phylogeny　106

| | | |
|---|---|---|
| 3.2 | Re-assessment of the incompleteness of the palaeontological record | 108 |
| 3.3 | The record of the Late Precambrian fauna applied to problems of phylogeny | 110 |
| 3.4 | Occupation of the marine environment: habitats and habits | 127 |
| 3.5 | Conclusions on the physical environments of the Late Precambrian Metazoa | 130 |
| 3.6 | The taxonomy of emerging diversity: the meaning of phyla | 132 |
| 3.7 | Rates of evolution | 135 |
| **4** | **The Precambrian–Cambrian transition** | **138** |
| 4.1 | Stratigraphic scales: boundaries and historical transitions | 138 |
| 4.2 | The fate of the Ediacarian faunas: extinction, survival, replacement | 144 |
| 4.3 | The Cambrian Period as the time of the first shelly fossils | 149 |
| 4.4 | The oldest Cambrian faunas and their evolutionary antecedents | 157 |
| 4.5 | Agglutinated, chitinous and mineralized body components in Ediacarian fossils and in some successors | 166 |
| 4.6 | Extrinsic and internal factors of biomineralization and its functional significance | 169 |
| 4.7 | Environmental changes at the Precambrian–Cambrian transition | 174 |
| 4.8 | Increase in diversity of trace fossils | 180 |
| 4.9 | Consequences and causes of evolutionary diversification | 188 |
| **5** | **Emerging animal life: thoughts on interactions of lithosphere, hydrosphere, atmosphere and biosphere** | **195** |
| 5.1 | Interactions of intrinsic and environmental factors in the emergent biosphere | 195 |
| 5.2 | Metazoan expansion in the marine biosphere: a three-stage process | 201 |
| 5.3 | Interactions in the light of plate tectonics | 207 |
| 5.4 | Animal life: past, present and future | 211 |
| | **Appendix: Questions of language and terminology** | 216 |
| | **References** | 218 |
| | **Dictionaries of scientific terms** | 237 |
| | **Author index** | 238 |
| | **Subject index** | 241 |

# Figures

| | | |
|---|---|---|
| 1.1 | Major divisions of earth history | 8 |
| 1.2 | Apparent timing of biospheric evolution | 9 |
| 1.3 | Comparison of prokaryotic and eukaryotic cells | 14 |
| 1.4 | The five kingdoms of organisms | 21 |
| 1.5 | Stromatolites | 23 |
| 1.6 | *Beltanelloides sorichevae* | 24 |
| 1.7 | Convoluted fecal string of a sediment-feeding metazoan | 24 |
| 1.8 | Worldwide occurrence of Ediacarian fossil Metazoa | 28 |
| 1.9 | Hypothetical stages in the phylogeny of the lower Metazoa | 34 |
| 1.10 | Depth distribution of benthic marine taxa in relation to deoxygenation of sediments | 38 |
| 1.11 | Outline of metazoan phylogeny | 41 |
| 2.1 | Map of fossil discoveries in the Flinders Ranges | 45 |
| 2.2 | Stratigraphic columns of Late Proterozoic in central Flinders Ranges | 46 |
| 2.3 | Typical exposure of fossiliferous Ediacara Member | 48 |
| 2.4 | Representative metazoan fossils from the Ediacara–Flinders Ranges area | 58–9 |
| 2.5 | Comparison of Ediacarian and living polychaete worms | 63 |
| 2.6 | *Bunyerichnus dalgarnoi* | 68 |
| 2.7 | *Planolites* sp., central Australia | 70 |
| 2.8 | *Charniodiscus?* sp., Officer Basin | 71 |
| 2.9 | *Arumberia banksi*, Amadeus Basin | 72 |
| 2.10 | Distribution of Nama rocks, Namibia | 75 |
| 2.11 | Stratigraphic column of Nama succession | 76 |
| 2.12 | *Cloudina hartmanae* | 78 |
| 2.13 | *Pteridinium simplex* | 79 |
| 2.14 | Reconstruction of *Ernietta* | 80 |
| 2.15 | *Ernietta plateauensis* and *Protechiurus edmondsi* | 82 |
| 2.16 | Fossil localities in Upper Vendian, northern Russia | 84 |

| | | |
|---|---|---|
| 2.17 | Stratigraphic columns showing fossiliferous strata | 85 |
| 2.18 | Selected fossils from the Ediacarian fauna of northern Russia | 86 |
| 2.19 | *Redkinia spinosa* | 88 |
| 2.20 | Tubes of Sabelliditidae | 89 |
| 2.21 | *Charnia* and *Glaessnerina* | 91 |
| 2.22 | Possible Charnian–Lower Cambrian succession in the Midlands area, England | 94 |
| 2.23 | Fossil localities in Upper Precambrian, southeastern Newfoundland | 95 |
| 2.24 | Stratigraphic column, Upper Precambrian, southeastern Newfoundland | 96 |
| 2.25 | Undescribed fossil from Mistaken Point, Newfoundland | 97 |
| 3.1 | Inferred relations of Hydrozoa, Anthozoa, Scyphozoa | 114 |
| 3.2 | Phylogenetic tree of fossil scyphozoans | 116 |
| 3.3 | Hypothetical stages in the phylogeny of the higher Metazoa | 120 |
| 3.4 | Hypothetical origin of arthropod legs and skeleton | 124 |
| 3.5 | Cross-section of trail, *Didymaulichnus miettensis* | 130 |
| 4.1 | Graph of ordinal diversity in Vendian and Lower Cambrian | 147 |
| 4.2 | Latest Precambrian shelly fossils | 152 |
| 4.3 | Lithofacies and biofacies in Hartshill Formation | 154 |
| 4.4 | Number of archaeocyathan genera plotted against time | 161 |
| 4.5 | Simple model of Cambrian transgressions | 177 |
| 4.6 | Diversity of Vendian and Cambrian metazoan families | 189 |
| 4.7 | History of familial diversity through Palaeozoic time | 193 |
| 5.1 | Relation between obliquity of ecliptic and ratio of insolation at poles to that at equator | 204 |

## Tables

| | | |
|---|---|---|
| 1 | List of fossils from Ediacara | 52–3 |
| 2 | Geographic distribution of Ediacarian metazoan taxa | 101 |
| 3 | Approximate time relations of latest Proterozoic glaciations and metazoan faunas, and fossiliferous Early Cambrian | 103 |
| 4 | 'Geochronostratic' scale of Precambrian–Cambrian transition | 141 |
| 5 | Provisional global correlation of the Ediacarian System | 145 |

# Preface

This book is a summary of the results of 25 years of work and lecturing. In 1979 I was invited by the State Bureau (now Ministry) of Geology of the Peoples Republic of China to tell staff and students in a number of colleges of my work. These lectures were grouped as a cycle of topics on Late Precambrian stratigraphy and palaeontology. On my return I decided that they could be usefully combined and expanded in the form of this book.

The book is intended for advanced students and for geologists and biologists interested in the early history of the living world and the marine habitat. It is not tailored to specific set courses of lectures given in certain years of specified courses of study; it does not represent transfigured lecture notes. It is meant to stimulate interdisciplinary studies; unavoidably, some parts of it may be too technical for some or too elementary for other readers. I have tried to make myself clear and I have provided or borrowed graphic tables where they may be helpful. A list of published dictionaries and glossaries of technical terms is added to the list of references.

Authors are frequently criticized for what they did not intend to write about. It may be appropriate to state at the outset that this book is neither a review of the literature, nor a comprehensive catalogue of descriptive data which can be found elsewhere; it does not deal with classification, nor with evolutionary, phylogenetic, or other biological theories which are at present too hotly debated to be handled without special care. Except where the context requires it, it does not try to answer criticism of what I have published, nor does it correct erroneous or misleading statements in other literature, simply because I wanted it to be less discursive and to keep it at a reasonable length. I have selected from the recent literature on geology, palaeontology and biology examples that seemed to me helpful in explaining relevant topics, and I have discussed the selected views of their authors in some detail. Many authors whose works I have read or with whom I have discussed various aspects of my studies are not mentioned, nor are those who have asked after my lectures significant questions that made me think again. Collectively, they all have my gratitude. If some ideas in this book are rightly theirs, I apologize. Some of mine may have drifted through normal scientific discourse into their writings and if this is the case, I am pleased. I have not claimed priority, except in my technical papers, and I have not spent much time investigating questions of who said what first. I am too old to gain or lose from such efforts. I have attempted to be fair and to say clearly and concisely what I think should be more widely known about my subject.

The book tries to describe and analyse, within the limits of my comprehension, the process of the appearance and diversification of animals in the biosphere. It is based on the study of documents of this process: fossils interpreted as functioning organisms of the past. The time frame in which this process took place has to be discussed and understood. Its span is about 1000 million years. The main factual evidence was collected worldwide during the last 25 years. My aim of describing this process, or a part of it, at such an early stage of its documentation is to show that it will be best understood as the result of complex interactions, rather than through *ad hoc* theories explaining parts of it by various unrelated causes acting at various times. It is too early to expect an elegant theory explaining how it all happened, but it is assuredly no longer true that we shall never know the essential steps in the process of major diversification of the animal world because it occurred so long ago.

It should not be thought that reference to close interactions is an easy way of letting other branches of science decide problems of palaeontology and evolution. In an unpublished address I said some 10 years ago: 'The whole story of relations between lithosphere, hydrosphere, atmosphere and biosphere is beginning to make sense now, not as four and not as two stories – inorganic and organic – but as one sequence of complex interactions.' And in J. G. C. Walker's work *Evolution of the atmosphere* (1977) which came to my knowledge after this book was written, he says: 'The goal has been to call attention to the possibility of change in the atmosphere and to illustrate the interrelationships of atmospheric evolution with the evolution of the crust, the oceans, and life'. The insistence on further study of interactions and the possibility of change is not an excuse for present ignorance but a call for future research programmes crossing boundaries of specialized fields.

# Preface

## Acknowledgements

I thank the University of Adelaide for hospitality and use of facilities offered after my retirement. I am grateful to the late Professors Sir Douglas Mawson and Arthur Alderman for encouragement in earlier days, to my friends Drs R. C. Sprigg, Mary Wade, W. V. Preiss and M. R. Walter and my colleagues, Dr R. J. F. Jenkins, Dr Brian Daily and Dr Brian McGowran, for their cooperation. Miss A. M. C. Swan assisted with preparation of illustrations and Mr R. Barrett with photographic work.

It will be obvious to readers that I – and perhaps they too – owe much gratitude to my wife Tina who patiently taught me Russian fifty years ago. I must thank her also for preventing my own premature fossilization.

Many workers on Precambrian palaeontology in many countries have been most helpful in providing material for inspection, sending publications or discussing problems. I regret that not all can be listed here but the following must be mentioned: Preston Cloud, Simon Conway Morris, Mikhail Fedonkin, Trevor Ford, Gerard Germs, Boris Keller, Igor Krylov, Vladimir Menner, Vladimir Missarzhevsky, Hans Pflug, Bill Schopf, Boris Sokolov. Special thanks are due to the US National Academy of Sciences, the Academy of Sciences of the USSR and the Ministry of Geology of the Peoples Republic of China for their invitations.

I am grateful to the following for permission to reproduce illustrations from their publications: The Editor, *Paleobiology* (Figs. 1.1, 1.2, 4.1, 4.6); W. H. Freeman & Company (Fig. 1.3); American Association for the Advancement of Science (Fig. 1.4 and Table 5); Verlag Paul Parey, Hamburg (Fig. 1.9); Chief Editor, *Zoologica Scripta* (Fig. 1.10A), Springer Verlag Heidelberg (Fig. 1.10B); The Editor, Geological Society of Australia (Fig. 2.1); The Editor, Royal Society of South Australia (Figs. 2.2 and 2.14); Springer Verlag New York Inc. (Figs 2.10 and 2.11); Geological Survey of Canada (Fig. 2.24); E. Schweizerbart'sche Verlagsbuchhandlung (Fig. 3.4); The Editor-in-Chief, Systematics Association (Figs. 3.3 and 4.5); Elsevier Scientific Publishing Co. (Fig. 4.3), Palaeontological Association (Fig. 4.4).

It is my pleasant duty to acknowledge the helpful cooperation and the efficient work on production of this book by Cambridge University Press.

# 1
# Precambrian life and its environment: a review

## 1.1 Theories about the origin of life and its environment

In the eighteenth century, James Hutton, one of the founders of the science of earth history, saw 'no vestige of a beginning, no prospect of an end'. Much has been learned in the 200 years which have elapsed since these words were written. We still see them as containing rather more than a grain of truth but we give them a different meaning from that which was in Hutton's eighteenth-century mind. Our aim now is to understand ourselves and our environment as part of an ongoing process of change. The model, still frequently if unconsciously applied because it is so close to the individual's experience, of birth, youth, maturity, old age and death, is not helpful when we consider superindividual processes in the universe. It does not imply a tidy and tedious calendar of a journey from point A, labelled 'beginning' to point B, 'the end', of which we wish to compile a partial diary. The aim of our science today is to compile a record of interactions which help us to understand the complexity of ourselves and our environment as revealed by ongoing research at ever increasing rates. The part of history of the universe which is our subject is animal life and its increasing diversity. To be sure, it had a beginning in the sense that there was a time when there were no animals, but, as Hutton recognized, the record contains no vestiges of a beginning and it would be a futile strategy to search for them. The history of life on earth may come to an end: we know that our technology has given us the means to end its existence but no end is in sight, at least at this point in time. James Hutton wrote that 'time is to nature endless and as nothing'. It has taken two centuries of thought and experiment to give content and definition to the concept of geological time which Hutton intuitively felt as 'endless' compared with experienced human life time and therefore as 'nothing' within our experience of time. The study of a process implies the study of rates of change in time and therefore the Huttonian endlessness must be measured and the supposed nothingness must be filled with evidence of events. We operate now with a geological time scale (Van Eysinga 1975, Cohee, Glaessner & Hedberg 1978). The order of magnitude of divisions of this scale, required for the study of the emergence and diversification of animals, is from 1 to 1000 million years (m.y.). The major division of the scale with which we are concerned is labelled Precambrian, for reasons which will be examined in detail in Chapter 4.

The origin of life is the result of circumstances and events close to the beginning of Precambrian time. In the absence of material vestiges of the first organisms we must briefly consider current theories about the beginning and about the distinguishing characters of what is called life. The question of the origin of life as a subject for scientific investigation was first raised in theoretical discussions independently by Haldane (1929) and Oparin (1924). The basis for a unified theory of the abiogenic origin of life is its basic chemical uniformity, together with well-founded assumptions about the possible composition of the atmosphere, hydrosphere and lithosphere at the relevant time, 3500–4500 m.y. ago. Haldane invoked from the then available experimental evidence the concept of an abundance of organic molecules ('sugars and some of the molecules from which proteins are built up') which are necessary for the formation of the elementary building materials of all organisms. He postulated that 'they must have accumulated till the primitive oceans reached the consistency of hot dilute soup' (Haldane in Bernal 1967, p. 246). This came to be referred to as the primitive soup or 'primordial broth'. Oparin considered the behaviour of colloidal solutions, pointing to the importance of cell walls and heterotrophic nutrition (from pre-existing organic compounds) in early organisms. Several years later, the experiments of Miller (1953) (see also Miller & Urey 1959) showed that energy discharge (electric sparks) in oxygen-free gas mixtures produced amino acids, the building stones of proteins, and similar experiments with other energy sources and gas mixtures produced further essential biological molecules.

We can now summarize the theoretical requirements for the origin of life as follows.

(1) The formation of organic molecules from which the basic components of living systems (proteins, lipids, carbohydrates, nucleic acids) can be built up. Some of their precursors occur in interstellar space. They have been identified by radioastronomy (see Chang 1981). Others occur in meteorites, particularly those belonging to the class of carbonaceous chondrites, or they could exist in comets. Some can be and have been produced experimentally from what would have been common materials in the

primitive earth crust and a reconstituted early atmosphere, in the absence of free oxygen.

(2) The atmosphere had to be reducing but the arguments continue about the need for a strongly or minimally reducing one, and accordingly about its composition. It may have contained hydrogen, water, ammonia, methane; or water, nitrogen and carbon dioxide; or carbon monoxide, hydrogen sulfide, ammonia and methane.

(3) A source of energy was required for the reaction of synthesis of prebiotic molecules. This could have been lightning which is essentially an electric spark, as in the Miller experiment, or ultraviolet radiation which was then unimpeded by the subsequently formed ozone screen in the upper atmosphere, or volcanic heat, as in hot springs, or solar heat, as in a desert where dew falls or where tide pools exist on an arid shore. Ionizing radiations have also been considered.

(4) 'The continued supply of external energy by radiation, etc., led to a build-up of substances of high free energy content on the early earth. The products must have remained dissolved or suspended in the water... The aqueous solution and suspension was the medium in which according to Oparin and Haldane life developed – Haldane's "dilute soup"' (Broda 1978, p. 27). No quantitative estimates of the possible dilution (or concentration) of this 'soup' are available. The need for an additional mechanism for localization of synthetic processes was felt by Bernal as early as 1951 when he suggested that some of the ubiquitous clay minerals, or alternatively quartz crystals, may have acted as catalysts for polymerizations. The difficulty of having an ocean turned into a soup bowl so as to bring organic molecules close enough together to make them react in quantity is only slightly mitigated by modern geological thought about the existence of many 'small' oceans in the early stages of crustal evolution. The unease about the organic-filled prebiotic ocean led to speculation about hot springs or lagoons as the geographical environment for the origin of life. However, with this model intercommunication of possible sites of reaction would be lost and the biochemical unity of the nascent life would be harder to understand.

(5) An unsolved problem is the beginning of the mechanisms confining the molecules required to form individual organisms. All living organisms form cells, except viruses which 'are not relevant to the origin of life' (Broda 1978, p. 30), being essentially adaptations of biochemical systems to parasitism. Cells are enclosed in membranes consisting of proteins and lipids. The cell membrane not only keeps together and separates from the environment the basic materials for life processes (metabolism, self-reproduction, bioenergetics) 'but is also the place of the mechanism for

transport into the cell and out of the cell, including the pumps for active transport' (Broda 1978, pp. 43–4). Protective cell walls consisting of sugars and peptides or of cellulose or chitin may surround the cell membrane but do not occur in primitive animal cells, with grave consequences for the possibility of their preservation in rocks. The recognition of the problem of separation of individual protobionts from the primeval broth led Oparin to fruitful experimentation with coacervates, self-congregating units in colloidal systems. It also led to preparation of 'microspheres' by Fox (see Fox & Dose 1972) and his followers. They are natural aggregations of proteinoids with interesting cell-like properties and appearance.

(6) The final problem is which – if any – of the processes of prebiotic chemical evolution, demonstrated to be possible by experiment, have actually led to the origin of life. There is some agreement about the actual occurrence of chemical evolution (Calvin 1969). It suggests a natural selection among chemical processes and products, leading to the survival of those most suited to continued self-reproduction through adaptation to the constraints and resources of the environment. This formulation points to two aspects of the problem which are under active investigation but still unsolved. Self-reproduction of all organisms is based on automatic transmission of information. This is contained in the genetic code, incorporated in the structure of the DNA molecule and uniform throughout the existing organic world. Its origin is still obscure. The second problem is the demonstration that what could have happened in the earliest stages of the history of the biosphere has actually happened. This would be possible if organic remains claimed to be produced by early organisms could be reliably distinguished as formerly living organisms from prebiotic chemical fossils.

The discovery of apparent fossil remains of single-celled organisms resembling bacteria and algae in rocks more than 3000 m.y. old (Barghoorn & Schopf 1966, Pflug 1966, Schopf & Barghoorn 1967) is significant. Critical reviews by Schopf (1975a) have thrown doubt on the status of these fossils as organisms and indicate their resemblance to shapes assumed by abiotic, in this instance probably prebiotic, organic matter. They have been referred to, with other similar finds, as 'fossil-like objects' (Schopf 1978). However, fossil sedimentary structures of organic origin known as stromatolites occur in rocks 3500–2500 m.y. old in Australia, southern Africa and Canada (Walter 1983). It is now well known that they are formed by the life activities of Prokaryota (bacteria and blue-green algae) (see Walter 1976, 1977, 1978, 1983). Undisputed fossil remains of microorganisms which produced stromatolites have been observed. There is evidence (Walter 1978, 1983) that some Archaean stromatolites were built by

*The origin of life and its environment*

photosynthetic Cyanophyta and by purple or green bacteria which assimilate carbon dioxide but do not produce free oxygen (Broda 1978). Many formerly existing forms of life, including microbes, must have been supplanted sooner or later by more efficient ones. This will become increasingly important as our story progresses but it causes the greatest difficulty in reconstructing microbial life of the past, a problem of classifying organisms whose observable life processes are vastly more significant than their morphological configurations.

Before we proceed to the next stage in the history of life it is necessary to consider the placing of the documented events in precise time scales and also to understand fossilization processes which convert the results of life processes into components of sedimentary rocks.

## 1.2 The Precambrian, its subdivisions, and the dating of biohistoric events

The construction and application of a geological time scale is almost as old as the science of geology. It implies a division of the uninterrupted flow of time during the entire history of the earth by a sequence of distinctive events. The model for it was, either consciously or unconsciously, human history with its division into antiquity, the Middle Ages and modern times, and its subdivisions such as Egyptian or Chinese dynasties, or the reigns of European monarchs, or socio-economic changes. There are also unique marker events like the Renaissance or the French Revolution, and the divisions of prehistory according to technological advances such as stone age, bronze age, iron age. The scales define primarily only *sequence* (the placing of other events as earlier or later than the key event), not *chronology* (dating of events and consequently duration of intervals). The geological chronology is being established with ever increasing accuracy by relating geological events to clock-like, fixed-rate processes such as radioactive decay. This makes it possible to calibrate the sequential stratigraphic scale. This scale was established early, by the application of the law of superposition to stratigraphic sequences. Except for later tectonic disturbances, lower strata must have been laid down before higher ones. The study of fossils contained in the rocks showed more than 150 years ago that they facilitated distinction between rocks occurring in sequence, and comparison of such sequences. These observations led soon to the development of biostratigraphy which eventually found in irreversible evolution and concomitant extinction of forms of life a guide for the recognition of increasingly finer divisions of the geological time scale. Biostratigraphy could be applied as far back in history as abundant fossils could be found, that is, to the beginning of the Cambrian

Period, the earliest part of the Palaeozoic Era. This and the two succeeding eras (Mesozoic and Cenozoic) were grouped together as the Phanerozoic Eon whose name was taken to mean the time of visible (or obvious) life.

When these names were proposed, no obvious traces of life were known from rocks older than those of Cambrian age, hence they were lumped together as Precambrian, of unknown duration. When calibration indicated that Precambrian time comprised about seven-eighths of geological time and when interest in Precambrian rocks increased for various other reasons, an extension of the more detailed time scale became necessary. The Subcommission on Precambrian Stratigraphy (a unit of the International Union of Geological Sciences) is preparing proposals which will eventually be submitted to an International Geological Congress for ratification. A proposal for a conventional division of Precambrian time into eons, the Archaean and the Proterozoic, has been approved by the subcommission. The boundary is defined by its chronometric (isotopic, commonly but loosely called radiometric) age, at 2500 m.y. ago. Possible subdivisions of the Archaean are still under discussion. The oldest presently known rocks are about 3800 m.y. old. Nothing is known from geological data about events between that age and the formation of the earth about 4600 m.y. ago. The dating of this event is based on cosmological evidence from meteorites and moon rocks. There is increasing information about the Archaean lithosphere, its composition and dynamics (Windley 1976, Tarling 1978). By the end of that eon, the earth's crust appears to have evolved significantly, to approximate the present lithosphere more closely, and so the concept of an eon boundary at about 2500 m.y. (James 1978), which is also significant for the history of life, is likely to find wide acceptance. A subdivision of the following eon, the Proterozoic, into three eras has also been proposed. The terms Early, Middle and Late Proterozoic will be used here for these divisions. Their proposed boundaries at 1600 and 900 m.y. (Sims 1980) are times of significant geological events documented in many continents.

The end of the Precambrian is necessarily the beginning of the Cambrian. Its precise definition is being considered by an international working group, which intends to base it on biostratigraphic changes in a stratotype section which has yet to be designated. This means that it will be based on the same principles as the Phanerozoic scale (Hedberg 1974) and that its age will be established by the dating of that concrete boundary event rather than by being based on an abstract chronological definition alone. The age of the base of the Cambrian is generally considered as approximately 550–570 m.y. The uncertainty may be greater than this range and further isotopic dating is urgently required. This boundary and its implications will be considered further in Chapter 4.

From the viewpoint of biostratigraphy, the Proterozoic is the time when stromatolites became abundant, while animal fossils from at least the greater part of this era are absent or rare. The middle of Precambrian time, between the age of the oldest rocks and 570 m.y. would be at about 1600 m.y. The informal concept of a mid-Precambrian time to which reference is made occasionally would encompass some early and some middle Proterozoic time from, say, about 1800 to 1400 m.y. ago. As we shall see, this is thought to be the time of the emergence of eukaryotic cells, with nuclei and organelles, from their prokaryote, non-nucleated predecessors. This important evolutionary step either rapidly led to or actually coincided with the first appearance of animal protists (Protozoa). The refinement of meiosis in cell division, sexual reproduction with its concomitant increase in evolutionary potential, colonial organisation and the origin of Metazoa, multicellular animals developing distinct tissues, followed sooner or later. The order of these events is based firmly on biological theory (see Margulis 1970, Patterson 1978) and will be further discussed in section 1.4. The palaeontological evidence and its geochronological dating are uncertain. This uncertainty encompasses nearly the entire mid-Precambrian interval, if not the entire Early and Middle Proterozoic, up to 1000 m.y. The evolutionary rates which determine the timing of these early stages of metazoan prehistory are unknown and cannot be known as long as remains of fossil animals, body and trace fossils, from sediments of Early and Middle Proterozoic age are unknown. Cloud (1976a) believes that the first Metazoa appeared about 700 m.y. ago which is the approximate age of the first unquestioned remains of metazoan body fossils. Older traces of animals exist but regrettably little is known about them. The first manifest metazoan fossil representatives are not what we would expect on theoretical grounds, i.e. small, planuloid or at the most advanced possible level, primitive coelenterates. It is more than likely that such animals could not have left traces in the sediments but a time span should and could be allotted to them without undue appeal to speculation. This has been attempted. The curve depicting oxygen accession to the atmosphere – oxygen being essential for oxidative respiration – is occasionally invoked as a means of dating biological events. However, the physical theory of oxygen (and ozone) accession, developed by Berkner & Marshall (1965) and brought to the attention of geologists by Cloud (1968), Fischer (1965, 1972) and Rutten (1962, 1966) who have modified it, does not result in a time scale. Berkner & Marshall placed the 'Pasteur point' (1% of the present atmospheric level (PAL) of oxygen) at the beginning of the Cambrian, which was thought to be the time of the oldest fossil animals. We know now that this is incorrect. Not enough is known quantitatively to calibrate precisely in terms of geological time the Berkner–Marshall curve of oxygen accession.

Cloud (1976a,b) divided the post-Archaean (Proterozoic) time, on geochemical grounds of change from unoxidized to oxidized epicratonal sediments and on grounds of biological change from Prokaryota to Eukaryota, into Proterophytic and Palaeophytic Eras, with a boundary at about 2000 m.y. (Figs. 1.1, 1.2). There is at present no indication of general acceptance of this biostratigraphic scale and nomenclature. Even soundly based and euphonious names are not necessarily generally accepted. Time scale divisions are conventions depending on consensus rather than on strictly scientific generalizations of a multitude of empirical data. International discussions on the time scale of the Precambrian and the naming of its subdivisions are planned for the 1980s. After a review of the Late

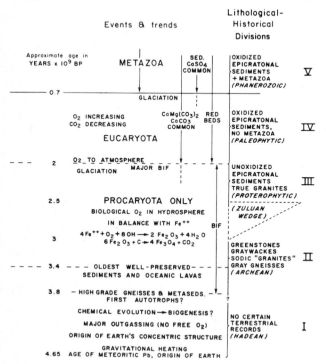

Fig. 1.1. Major divisions of earth history and related aspects of biological and geochemical evolution. (BIF – banded iron formations; BP – before present.) Adapted from Cloud (1976b).

Proterozoic Metazoa, the time framework for that period will be re-examined (Chapter 4).

## 1.3 The preservation and recognition of organic remains in rocks: chemofossils, fossil organisms, pseudofossils

Before proceeding with the study of the history of life to the point in time when animal remains become recognizable, we have to consider the nature of existing documentation of the life of the past and the interpretation of these documents. There is an analogy between them and the material evidence of human history. The analogy is valid in the sense that as we go back to the oldest times, the record of both kinds of history becomes more difficult to decipher and to interpret. It also becomes inherently more incomplete, full of gaps and false clues which are apt to mislead. Much has been written about the inherent incompleteness of the Phanerozoic fossil record. Estimates have been made of the percentage of living species and higher taxa which could not have been preserved. Much can be learned from the well-known Phanerozoic record about fossilization

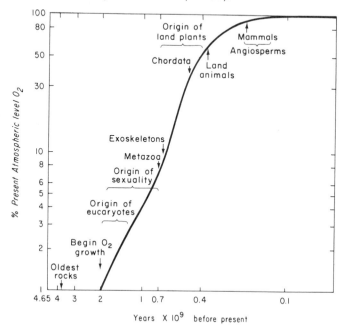

Fig. 1.2. Apparent timing of events in biospheric evolution compared with hypothetical levels of atmospheric oxygen. Both scales logarithmic; curve highly generalized, excursions from the mean likely. From Cloud (1976*b*).

potential, but as will be seen in Chapter 4, evolution of mineralized skeletons has made a quantum jump at the Precambrian–Cambrian transition. Their absence has not rendered the Precambrian entirely unfossiliferous as had been expected earlier, but it has necessitated much effort in building up a specific methodology of discovery and interpretation of Precambrian fossils which is still incomplete. The availability of fossils for the study of life history is a function not only of the fossilization potential of the biota of different periods but also of the exploration effort. In Darwin's time so few Phanerozoic fossils had been discovered that he was unable to make any significant use of them for his theories of evolution by natural selection. One hundred years later, no theoretical discussion of the course of evolution is complete without reference to palaeontological evidence. The rapid increase in the exploration and study of Precambrian geology, stimulated by the abundance of mineral resources formed during that time span, and the recognition of its great length, has revealed an abundance of Precambrian fossils. In general terms they are quantitatively and qualitatively inferior to those of later Phanerozoic age but they are highly significant for several reasons. It is true that with greater age of rocks there are greater chances of their alteration by repeated tectonic and magmatic processes leading to the destruction of organic remains but this is not the general fate of all Precambrian rocks, contrary to what had been expected. Exploration has revealed that many of them have remained not too strongly altered to contain organic remains. There are thick sequences of sediments previously considered as unfossiliferous (and often in the absence of clear evidence dated as Palaeozoic) which, with the application of new biostratigraphic methods, have been shown to be of Precambrian age and which have since been dated chronometrically. New methods of the physical sciences, electron microscopy and modern analytical chemistry, are leading to discoveries of great significance which are still being evaluated. The occurrence of organic matter in ancient rocks has been known for a long time and has proved puzzling, in the absence of other traces of fossil organisms and in the face of authoritative views of their presumed absence. Molecular biology, a dominant branch or viewpoint in biological sciences, has led to suggestions of a 'molecular palaeontology' (Zuckerkandl & Pauling 1962, Zuckerkandl 1965, Florkin 1966, McLaughlin & Dayhoff 1973, Broda 1978). They are based on evolutionary studies of proteins (haemoglobins, cytochromes) from mainly microbes and vertebrates, and evaluation of their genetically controlled alterations with respect to phylogeny and time. Biochemists have considered chemical evolution (Calvin 1969). The evolution of bioenergetic processes (Broda 1978) is not a supplement but an essential substrate for these studies, with far-reaching consequences for the understanding of the first appearance of animals whose metabolic activities and bioenergetic requirements are different from those of their predecessors.

The palaeobiological significance of the study of stable organic compounds isolated from ancient rocks as 'chemical fossils' has been recognized in organic geochemistry (e.g. McKirdy 1974). The main problems are the possibility of abiotic or prebiotic origin of some organic compounds, the danger of contamination when the analyst has to deal with minute quantities, the possibility of migration of fluids through even weakly porous or fractured rocks during long periods of time, and the alteration of chemical structures by diagenesis and incipient metamorphism. In recent years there has been much progress in the understanding of the alteration of insoluble organic matter in sedimentary rocks (kerogen) which must be syngenetic and coeval with the enclosing rock. The possible linking of chemical marker fossils with evolutionary events is particularly significant. The early occurrence of photosynthesis is partly deduced from such studies but the dating by chemofossils of the change from prokaryote to eukaryote cells or from fermentation to oxidative respiration is still essentially in the future.

The recognition of fossil organic remains on the basis of their configuration rather than their chemical composition is particularly difficult in ancient rocks. These difficulties have been discussed by Hofmann (1971, 1972) who proposed an elaborate classification which distinguishes 'pseudofossils' and 'problematica' or 'dubiofossils' from definite remains of organisms ('fossils'). Here we shall distinguish between difficulties of recognition caused by (1) alteration of rocks, (2) similarities of organic and inorganic configurations and (3) problems of evolutionary morphology affecting fossilization.

(1) Alteration. It is generally true that Precambrian rocks are more highly altered, i.e. more affected by diagenetic, metamorphic and tectonic processes than younger rocks. Fossils have been found in younger metamorphic rocks but their preservation under conditions of elevated temperatures and pressures or metasomatic processes is dependent on a degree of robustness of embedded organic remains which was not generally attained, at least by animal bodies, until Phanerozoic time. The search for remains of fossil organisms is generally confined to special classes of altered sedimentary rocks. Early diagenesis can enclose organic remains so as to seal them hermetically against further alteration. This can occur in a matrix of chert (cryptocrystalline silica) or in films of organic matter (kerogen). It may also be possible to recognize traces of organic activity (burrowing, coprolitic pellets) in the fabric of mildly altered sediments.

Few if any of them will be unquestionably biogenic and most of them will be therefore classed as 'problematica'. The odd shapes of the results of some tectonic deformations (boudinage) and of concretions will generally make them recognizable as pseudofossils.

(2) Similarities of organic and inorganic configurations. Geometric regularity of configurations which could not be immediately identified as crystals was taken as an indication of organic origin, at least in the early stages of the search for Precambrian fossils. Clay pellets often resemble fossil shells, infillings of drying cracks can resemble worm burrows or bodies, the outlines of angular shards of dried clay layers have been combined to produce misleading reconstructions of imaginary fossil arthropods. Fractured or sectioned glass bubbles in volcanic tuffs may strikingly resemble sponge spicules or chambered Foraminifera or Radiolaria. Configurations such as vertical stalks ending in lobate structures have turned out to be sand volcanoes. Jellyfish-like concentric structures on bedding planes may be found, on closer examination, to have vertical feeder channels and to be formed by escaping gas or fluids. Others, without stalks, have turned out to be pressure marks or moulds of crystal rosettes. The possibility of mechanical origin such as current effects or bedding lineations must be rigorously excluded or at least made highly improbable by observation of specific characteristics before similar configurations are admitted as genuine fossils. An interesting example of the need for caution in evaluating the biogenicity of reputed Precambrian fossils arises from a comparison of an early distribution table of groups of fossils in the Precambrian (Glaessner 1966, Fig. 2) with present knowledge. Ten stratigraphic units from various countries were included as containing either trace fossils or (animal) megafossils. Two of them were considered as questionably fossiliferous. Three supposed trace fossil occurrences and the two questionable ones are now known to contain only configurations of mechanical origin, one contains megascopic algae mistaken for animal remains. Some trace fossils from the Precambrian–Cambrian Vindhyan rocks of India may be Cambrian and one figured trace was apparently caused by the rasping of the radula of a living snail on the surface of the rock. The origins of other trace fossils are still being questioned. Cloud (1968, 1973) has acted consistently as an invaluable sceptic in the evaluation of Precambrian configurations considered as organic remains. They require continued strenuous efforts to verify their organic origin, more than do younger fossils.

(3) Evolutionary problems. Not long ago the course of evolution before the Cambrian was undocumented and even the fauna of the Early Cambrian was poorly known. This explains why the occurrence in the

Precambrian of large eurypterids ('reconstructed' from the outlines of large clay flakes) or of articulate brachiopods (which subsequently proved to come from inliers of Cambrian rocks in Precambrian terrains) was accepted. True shells or skeletons originated about the time of the transition from Precambrian to Cambrian (see Chapter 4). The real problem in finding and interpreting Precambrian fossils is the fact that these animals were without shells or skeletons (soft-bodied). The preservation of soft-bodied organisms poses many problems of the reconstruction of their original form and of the taxonomic assessment of their variability. These problems and the methods assisting in their solution are rarely discussed in the literature. A recent authoritative statement on fossilization (A. H. Müller 1979) devotes one paragraph to the preservation of soft parts in some 70 pages of text. Meanwhile the wasteful labour of description and publication of pseudofossils mistaken for fossils continues.

## 1.4 The cell: organization and life processes. The 'kingdoms' of organisms

All organisms, except viruses, are built up from cells or organized as cells. They are separated from their environment and able to communicate with it through the structure of their surrounding membranes. To fulfil their tasks, including the basic one of self-reproduction, the cells cannot be simple pieces of bioengineering. The simplest kind, a bacterial cell, has been compared in complexity with a chemical factory, with an information content, at the atomic level, of $10^{12}$ 'bits' or 'yes/no decisions' built into it. There are among living organisms two basic kinds or levels of cell organization, the prokaryote cells of the bacteria and the blue-green algae (which are now frequently referred to as cyanobacteria), and the eukaryote cells which make up all other organisms. The prokaryote cells are mostly small (1–10 $\mu$m), without a nuclear membrane or endoplasmic reticulum (cytoplasmic membrane system), no mitochondria (enzyme-containing bodies), no chloroplasts for photosynthesis; they divide by binary fission (Fig. 1.3). The eukaryote cells are generally larger (10–100 $\mu$m, exceptionally larger by up to three orders of magnitude) with a nuclear membrane and with the organelles which are absent in the prokaryotes; cell division occurs by mitosis followed by meiosis, and the cells have complexly but uniformly structured flagella or cilia. They are primarily aerobic and produce oxygen in the process of photosynthesis. 'The numerous and fundamental differences between eukaryotic and prokaryotic organisms...have been fully recognized only in the past few years. In fact, this basic divergence in cellular structure which separates the bacteria and the blue-green algae from all other cellular organisms, probably represents the greatest single

# 1: *Precambrian life and environment: a review*

Fig. 1.3. Comparison of prokaryote and eukaryote cells. The drawings represent a bacterium compared with a single-celled green alga. Its cell contents include ribosomes and food storage bodies (unlabelled). The chloroplasts are also known as plastids and the stigma as the eye spot (Tappan 1980, pp. 895–6). From Schopf (1978). Copyright © 1978 by Scientific American Inc. All rights reserved.

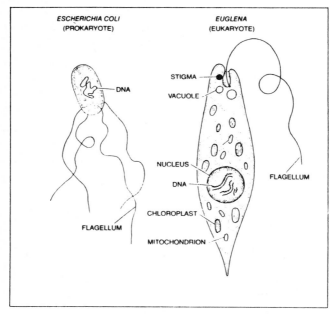

| | PROKARYOTES | EUKARYOTES |
|---|---|---|
| ORGANISMS REPRESENTED | BACTERIA AND CYANOBACTERIA | PROTISTS, FUNGI, PLANTS AND ANIMALS |
| CELL SIZE | SMALL, GENERALLY 1 TO 10 MICROMETRES | LARGE, GENERALLY 10 TO 100 MICROMETRES |
| METABOLISM AND PHOTOSYNTHESIS | ANAEROBIC OR AEROBIC | AEROBIC |
| MOTILITY | NONMOTILE OR WITH FLAGELLA MADE OF THE PROTEIN FLAGELLIN | USUALLY MOTILE. CILIA OR FLAGELLA CONSTRUCTED OF MICROTUBULES |
| CELL WALLS | OF CHARACTERISTIC SUGARS AND PEPTIDES | OF CELLULOSE OR CHITIN, BUT LACKING IN ANIMALS |
| ORGANELLES | NO MEMBRANE-BOUNDED ORGANELLES | MITOCHONDRIA AND CHLOROPLASTS |
| GENETIC ORGANIZATION | LOOP OF DNA IN CYTOPLASM | DNA ORGANIZED IN CHROMOSOMES AND BOUNDED BY NUCLEAR MEMBRANE |
| REPRODUCTION | BY BINARY FISSION | BY MITOSIS OR MEIOSIS |
| CELLULAR ORGANIZATION | MAINLY UNICELLULAR | MAINLY MULTICELLULAR, WITH DIFFERENTIATION OF CELLS |

evolutionary discontinuity to be found in the present-day living world' (Stanier, Adelberg & Doudoroff 1963). The most unexpected fact observed in Precambrian palaeontology, the dominance of remains of bacteria and blue-green algae in rocks dated as representing the first half of Precambrian time, is an expression of this first and highly significant 'evolutionary discontinuity'. It raises three questions: (1) What evolutionary step or steps could have bridged this discontinuity? (2) How are these evolutionary events related to changes in the environment? (3) What evolution, if any, took place at prokaryote level, during Early Precambrian time, through more than 2000 m.y. since deposition of the oldest fossiliferous rocks?

(1) One perhaps surprising answer to the first question was given by the symbiotic theory of the origin of the eukaryotes (Margulis 1970). Margulis considers the classical view that the more highly organized forms of life originated from bacteria and blue-green algae 'by the accumulation of selectively advantageous mutations [as] inconsistent with many facts'. The symbiotic theory asserts that prokaryotes acquired such organelles as mitochondria in the form of symbiotic aerobic bacteria; flagella and cilia were similarly pre-existent as spirochaetes and incorporated in prokaryotes where they evolved eventually to participate not only in locomotion but also in advanced reproductive processes ('mitotic apparatus'). Some blue-green algae were originally endosymbiotic in protozoan-like heterotrophs to form plastids and finally chloroplasts as sites of photosynthesis. The details of this theory and the factual evidence need not concern us here. They were lucidly presented by Margulis, with due consideration of available palaeontological and geohistorical data documenting as far as possible what could have happened at the relevant time in the history of life. G. E. Hutchinson remarked in his foreword to Margulis' work (1970, p. xvii) that 'not every interested biologist may accept all her suggestions'. This has proved correct (see Broda 1978, pp. 132, 180 for contrasting views; Taylor 1974, with reply by Margulis 1975; Cavalier-Smith 1975, discussed in detail by Cloud 1976a). A review of relevant new observations and interpretations appeared too late to be considered here (L. Margulis, *Symbiosis in cell evolution*, W. H. Freeman & Co., San Francisco, 1981). It strengthens and widens the basis for the endosymbiosis theory. At least some of its postulates have proved widely acceptable, fruitful for further research, and significant for the origin of animals, an event not likely to have occurred much later than the first appearance of eukaryotic cells in mid-Precambrian time. The various hypotheses about their evolution and its timing were clearly and fairly reviewed by Tappan (1980, pp. 84–96).

(2) The answer to the second question is that these evolutionary events at cell level do not presuppose any rapid and fundamental changes in the

environment. It may be said that they have gradually laid the foundations necessary for an expansion of the biosphere in mass, diversity and in its influence on the history of the earth. The lithosphere and hydrosphere of at least Early Proterozoic time were, according to geochemical and tectonic data, probably not very different from their present composition and dynamic state, though not as rich in diverse ecological sites (niches) for various life forms. The most distinctive development in the physical environment in Early Precambrian time was the accumulation in the atmosphere of free oxygen. According to Cloud (1976a) it amounted to about 3% PAL at the time of the origin of the eukaryotes. There is biogeochemical evidence of photosynthesis occurring at the time of the formation of the oldest sedimentary rocks. It is still practised by a few bacteria. They never produce free oxygen but the blue-green algae and the plants do. The oxygen in the present atmosphere is the product of this life activity, with only minor contributions from photodissociation of water. The process is efficient enough: 'It can be calculated, for example, that the offspring of one gram of algae (assuming unlimited space and nutrients and a reasonable rate of cell division) could photosynthetically produce an amount of oxygen equal to that of the present atmosphere in less than 40 days' (Schopf 1975b, p. 55). But of course space and nutrients were not unlimited and so it took longer. How much longer is still a controversial question; estimates vary from the assumption of a fully oxygenic atmosphere in Early Precambrian time, which conflicts with the fact that unoxidized minerals of that age were found where they would have been exposed to the atmosphere when deposited, to Cloud's estimate (1976a) of only about 50% PAL at about the end of Cambrian time. Cloud has repeatedly drawn attention to a relevant geological fact. Banded iron formations, presently the major source of industrial iron production, had their worldwide maximum development about 2000 m.y. ago and then they disappeared from the record almost completely. They were formed under conditions different from those which the fully oxidized redbeds required for their deposition. Typically their colour is due mainly to quartz grains coated with iron oxides. Cloud considers this change as the first indication of conditions which permitted a change of life processes from fermentation to oxidative respiration, a change which now occurs in facultatively aerobic organisms at about 1% PAL in the organisms' environment (the so-called Pasteur point). This change in the organisms' energy acquisition occurred certainly under considerable selection pressure, as oxidative respiration is more than twice as efficient than fermentation, in terms of free energy produced by the basic reaction.

(3) The fact that Prokaryota (bacteria and blue-green algae) dominate

the Precambrian fossil record does not necessarily indicate very low rates of evolution during this long time. Studies of detailed stratigraphic sequences are still too few and too new for final conclusions to be drawn. Not all difficulties of interpretation of fossilization processes and of their influence on original micromorphology have been overcome. There are indications of evolutionary changes in size and complexity (Schopf 1977). Schopf has repeatedly stressed the importance of the development of mitotic and meiotic reproductive mechanisms of the eukaryote cell for significant increases in diversification rates. This is very likely but the palaeontological evidence for their occurrence is by no means unequivocal, because of problems in interpreting detailed observations on fossil remains of single or possibly dividing cells. What is probably more important is the need for the evolution of complex and sophisticated new biochemical pathways within the structural framework of the prokaryote cell to enable it to make efficient use of the changed environment for its metabolic and bioenergetic needs. The time required for this biochemical evolution under conditions of a significant increase in the oxygen content of the atmosphere and the hydrosphere cannot be estimated but it was probably long in geological terms. In other words, the morphological simplicity of the fossil and existing prokaryotes may conceal a lengthy evolution of their basic life activities and adaptations and possibly also extinctions of less well adapted life forms.

Energy in the form of light is a precondition of life which (to use a profound thought expressed variously at different times by such eminent physical scientists as Boltzmann, Schrödinger or Broda) interposed itself into the flux of radiant energy from the hot sun to the warm earth and on to cold space. Oxygen is the basis for oxidative respiration, the efficient transformation of radiant energy to the energy of the chemical bond in the adenosine triphosphate (ATP) molecule of living organisms. The accession of oxygen to the atmosphere as the result of photosynthetic use of radiant energy is favourable for the further evolution of life under two conditions. Some of it had to be available to protect living organisms from harmful ultraviolet radiation, by forming an ozone screen in the atmosphere. The ability of some prokaryote cells to repair radiation damage has been proved experimentally but could hardly have been effective enough to assure their dominance during 2000 m.y. of early history of the biosphere. Other protective mechanisms have been invoked, such as the shielding effect of 10 m of water or of the calcareous deposits on algal mats which are preserved throughout the geological record as stromatolites. There have been objections to the presumed efficacy of these mechanisms. On the other hand, there have been objections to the suggestion that the ozone screen began to operate as soon as there was free oxygen in the atmosphere. If it appeared late, the apparent delay in the occurrence of abundant life on dry land could be explained. Abundant plant or animal fossils are not found in non-marine sediments older than 400–500 m.y. However, there may have been other reasons for the late appearance of abundant terrestrial life. The conclusion must be that the dating of the screening out of lethal ultraviolet radiation is still controversial: 600–650 m.y. ago according to Cloud (1968), or the time when as little as 0·1% PAL of oxygen was present (Ratner & Walker 1972).

The second precondition for the use of free oxygen in life processes is avoidance of its toxicity. While some biologists believe that the necessary protective enzyme systems could have evolved in a geologically insignificant time span, concomitantly with the respiratory pathways, Cloud (1976a) assumes a slow start of respiration in oxygen-poor environments and designates the onset of redbed sedimentation about 2000 m.y. ago as the time 'when enzymatic mediation of that particular atmospheric pollutant became efficient enough to tolerate $O_2$ levels above $\sim 1\%$ P.A.L.'

We can now turn to the question of the time of the first appearance of eukaryote organisms. Cloud (1976a), who has made the most comprehensive studies of the beginnings of biospheric evolution, concluded that the origin of eukaryotes may have occurred at any time from 2000 m.y. ago onward but that the oldest really persuasive morphological evidence for that development is not found until about 1300 m.y. ago. Before discussing the evidence for the earliest occurrence of animal remains in the geological record, the distinction between animals and other forms of life has to be considered. All animals and plants (all organisms other than bacteria and blue-green algae) have eukaryotic cells. Some exist as single cells or as colonial cell aggregates. They are known as Protista, or if they are known to live as animals do, as Protozoa. The multicellular animals with cell systems differentiated and functioning as tissues are known as Metazoa. It is the almost generally held view that the Metazoa evolved from the Protista. Within the living Protista the distinction between plants and animals is not significant but to some extent only a matter of terminology or semantics. There are single-cell organisms that sometimes nourish themselves by photosynthesis, and at other times swim about digesting food particles (Broda 1978, p. 138, with reference to Margulis 1970). Apart from those Protista which can function either as plants or as animals, photosynthesis in plants and phagotrophy in animals, which consequently develop locomotion, are generally valid distinctions. They are not important enough to exclude, for example, Phytomonadina (plant flagellates) from discussion in a standard textbook on protozoology (Grell 1973) and they could not generally be applied to extinct Protista but only to living organisms whose feeding habits can be observed. The old taxonomic term Protophyta has become obsolete; yet it does not make sense biologically to remove the photosynthetic flagellates from their systematic grouping with other algae as long as even the prokaryote Cyanophyta are commonly referred to as blue-green algae (Hanson 1977). The problem concerning us is the tracing of possible evolutionary pathways from early eukaryotes to Metazoa. The extent to which fossil remains can be used to test the historical reality of the theoretical conclusions concerning animal ancestors will depend on our ability to recognize fossil evidence of the distinctive animal *functions*: locomotion, and ingestion of food particles, with its corollary, the excretion of structured fecal matter. This is possible only at the grade of Metazoa. The recognition of fossil Protozoa as their possible ancestors is excluded in most instances by the difficulties of their preservation. While plant microbes have cell walls of cellulose, animal cells have less resistant membranes consisting of proteins and lipids, or chitin. Hardening of walls by agglutination of foreign particles, mineralization with silica or carbonates, or sclerotization of chitin are secondary phenomena which did not occur before Late Precambrian time.

The decay of the cell content during fossilization can leave a residue looking like intracellular membranes or organelles and a dark spot near its centre can resemble a nucleus. This problem has been investigated by experimental 'fossilization' of known organisms (Oehler 1976, Francis, Margulis & Barghoorn 1978), with different results according to the methods used. Similar difficulties cloud the interpretation of fossil evidence of cell division in eukaryotes as distinct from binary fission of prokaryotes, and much other palaeocytological detail (Knoll & Barghoorn 1975, Oehler, Oehler & Muir 1976). A significant conclusion from experimental testing of earlier observations states: 'Although it is a logical conclusion that the eukaryotic level of cell organization arose before 680 m.y. ago, there is no definite cytological evidence for an earlier date of appearance of eukaryotes in the fossil record'. (Francis *et al.* 1978, p. 97). On this negative evidence we cannot disregard 'logical conclusions'. Work on morphology, palaeobiochemistry and fossilization of Early and Middle Proterozoic microfossils is progressing and the latest conclusions about the timing of evolutionary and environmental events may be subject to further revision in the light of new data and advances in methods of study and experimentation. The richest and best preserved assemblages of such microfossils are from the Gunflint and Bitter Springs stromatolitic and non-stromatolitic cherts of North America and Australia, respectively. The first named is about 2000 and the second about 850 m.y. old and there are other similarly significant microfossils of intermediate age which have been discussed by Cloud, Schopf and others in numerous review articles. A relevant and cautiously expressed conclusion seems to be that the time of deposition of the Gunflint chert 'corresponds approximately to the transition from an oxygen-poor atmosphere to one in which there were significant levels of free oxygen', about 1800 to 2000 m.y. ago; a critical period in the history of life during which 'enhanced morphological experimentation produced the problematical microorganisms now found in the Gunflint' (Awramik & Barghoorn 1977, p. 128). While according to these authors the Gunflint microbiota is wholly prokaryotic, the Bitter Springs microbiota is likely to contain eukaryotes which may have first appeared 1400–1500 m.y. ago (Schopf & Oehler 1976). Comparison of early eukaryotes suggests affinities to green algae to most observers, but others have different opinions (see Knoll & Golubic 1979) or refrain from definite taxonomic assignments. The earliest microfossils resembling possible Protozoa are from the Chuar Group of the Grand Canyon, Arizona, dated at more than 650 and less than 850 m.y. old (Bloeser et al. 1977), from southwest Brazil (Fairchild, Barbour & Haralyi 1978), from Saudi Arabia (Binda & Bokhari 1980) and from the uppermost Riphean of Greenland (Vidal 1979) and Sweden (Knoll & Vidal 1980). The finds have been correlated with rock sequences which are about 700–800 m.y. old. These microfossils are described as Chitinozoa ('heterotrophic protists or primitive metazoans') by Bloeser et al., but as 'Chitinozoan-like' by Vidal and Binda & Bokhari.

We have seen that the simplest eukaryotes are the unicellular (or colonial unicellular) Protista and that among them the first manifestations of animal life must have occurred. They are unlikely to have left any recognizable structural marks on fossils or in rocks. The proposers of the multikingdom concepts of classification including a kingdom Protista have recognized this as a confederation of those eukaryotes which are neither Metaphyta nor Metazoa (Whittaker 1969, Whittaker & Margulis 1978). Living Protozoa which could serve as models for the first animals must be among the zooflagellates and ciliates. They are not likely to be found among the Rhizopoda of which those with preservable skeletons (Radiolaria, Foraminifera) appear late in the geological record, with clearly 'primitive' representatives initiating a comparatively well documented, complexly radiating evolution not earlier than in Phanerozoic time. Hanson (1977) considers the Protozoa as polyphyletic, i.e. as derived from various colourless algal protists. His conclusions on the origin of the Metazoa will be discussed together with others in Chapter 3. Here we state only that it is not unlikely that groups of Protista which could have included ancestors of Metazoa have survived to the present. If not, if these ancestors have become extinct, it is unlikely that we shall obtain any information about them from the geological record. In any case, what occurred among the Precambrian Protista was probably not so much a morphological change as one towards dominant ingestive nutrition. This form of nutrition combined with motility is found in phytoflagellates which are also equipped for photosynthesis. Hence evolutionary transition to animal protists could have occurred through loss of photosynthetic organelles (plastids).

Fig. 1.4. The five kingdoms of organisms. After Whittaker (1969, Fig. 3, simplified). (Copyright 1969 by the American Association for the Advancement of Science.) Numbered subdivisions, more or less relevant to discussions in the text, are as follows. Monera (or Prokaryota) – 1, bacteria; 2, Cyanophyta (or Cyanobacteria). Protista – 1, Ciliata (or Ciliophora); 2, Sarcodina (Foraminiferida, Radiolaria, amoebae); 3, Zoomastigina (animal flagellates). Plantae – 1, Rhodophyta; 2, Chlorophyta; 3, Charophyta; 4, Phaeophyta; 5, Bryophyta; 6, Tracheophyta. (Alternative classifications (Whittaker & Margulis 1978) include 1–4 and the lower Fungi with the Protista, as 'protoctista'.) Animalia – 1, Mesozoa; 2, Porifera (or Parazoa); 3, Coelenterata (Cnidaria and Ctenophora); 4, Platyhelminthes; 5, Aschelminthes; 6, Tentaculata (or Lophophorata); 7, Chaetognatha; 8, Annelida; 9, Mollusca; 10, Arthropoda; 11, Echinodermata; 12, Chordata. (In the later classification (Whittaker & Margulis 1978) the Pogonophora and Hemichordata were grouped with the Chaetognatha between 11 and 12.)

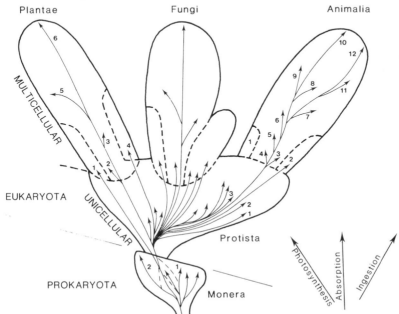

The desirability of excluding in a formal classification scheme the Protista from both the plant and animal kingdoms is obvious, despite its inconvenience. This, as Whittaker (1969, p. 154) remarks, 'is not so much the fault of the systematist as faults of the living world as a subject of classification'. The stream of evolution widens and divides, diversifying to fill available niches (Fig. 1.4). The level of organization above that of Protista is characterized by multicellularity and tissue development and the basic divisions are defined on the basis of nutrition because of its far-reaching consequences on structure and function. Nutrition is absorptive in Fungi. Therefore, and because of structural adaptations to their mode of life, they deserve separation from the other kingdoms. In the kingdom Plantae *sensu stricto* it is mainly photosynthetic, with nutrients obtained from water, 'soil' and air. The organisms are basically autotrophic and sessile and should include the multicellular 'higher' algae. The fact that nutrition is by ingestion of captured food in the kingdom Animalia raises the multifarious problems of locomotion. It has been suggested that with the Metazoa this kingdom should also include the small groups of Mesozoa and Parazoa, with absent or limited tissue differentiation. The Mesozoa are without fossil representatives and will not be considered here. The Parazoa (sponges) are here included with the Metazoa as animals. For the rest, widely accepted formal classifications of the kingdom Animalia will be used in this study. It is not its aim to criticize or develop them.

## 1.5 Proterozoic fossils and environments

At the beginning of the Proterozoic Eon, autotrophic prokaryotes (bacteria and blue-green algae) were in existence. The most striking and generally characteristic feature of the Early Proterozoic fossil record is the first abundant and widespread occurrence of stromatolites. It is related to and probably partly the cause of the greater abundance of limestone and dolomite deposition in shallow water. The first result of systematic studies of Precambrian stromatolites (Fig. 1.5) was the recognition of their biostratigraphic potential. Early Proterozoic (Aphebian), Middle Proterozoic (Early and Middle Riphean, stromatolite assemblages I and II–III, respectively) and Late Proterozoic (assemblages III and IV) were distinguished (see articles by Semikhatov, Preiss and Donaldson in Walter 1976, Chapter 7). The numbered assemblages (specified by Preiss in a note on p. 368 of his article) were described first in the USSR and later found in other regions. The subsequent studies of Aphebian stromatolites revealed considerable similarities in their configuration on the level of form genera but also sufficient specific structural differences to confirm rather than negate their stratigraphic potential, provided that their study was carried out in necessary detail. Due consideration had to be given to environmental, preservational and diagenetic factors. At the same time, considerable efforts were made to find and describe the fossil remains of the algal communities which had contributed to the construction of stromatolites (Schopf et al. 1977, Schopf & Prasad 1978) as well as non-stromatolitic occurrences of Proterozoic fossil algae. It was reported that about 20 distinctive assemblages of microfossils occur in biostratigraphically useful sequences, about 10 of them in strata which are more than 1000 m.y. old.

The abundance and diversity of stromatolites declined sharply in Late Precambrian time, from about 800 m.y. ago onwards (Gebelein in Walter 1976). The fact that they had flourished until that time points to the absence or insignificance of Metazoa and Metaphyta in their environment (photic, subtidal to intertidal) in which shelf-type limestones have been deposited. In subsequent times the predation by Metazoa together with competition by red and green algae and other marine plants confined the stromatolite builders mostly to hypersaline and freshwater environments. This occurred rapidly during Early Palaeozoic time and continued to the present when they survive as relicts.

The major primary producers of food resources for heterotrophic

Fig. 1.5. Stromatolites. *Gymnosolen ramsayi* Steinmann, from a boulder in the Late Proterozoic Tapley Hill Formation, Flinders Ranges, South Australia. Photo by courtesy of Dr W. V. Preiss.

24    1: *Precambrian life and environment: a review*

organisms in the sea are the photoautotrophic phytoplankton organisms. It is difficult to establish their fossil record through the great time spans of the Proterozoic. It has been suggested that their quantitative fluctuations in Phanerozoic times had a decisive influence on the evolution of the marine fauna (Tappan & Loeblich 1971). The present level of our knowledge does not permit reliable extrapolation into Precambrian time, for a number of reasons. Planktonic algae belonging to extant taxa make their first appearance in Phanerozoic strata: the chrysophycean coccolithophorids

Fig. 1.6. *Beltanelloides sorichevae* Sokolov (= *Beltanelliformis brunsae* Menner), from a bore at Leino, eastern Russian Platform, Vendian, Redkino Series. From Sokolov (1972*b*). Magnification: × 3.

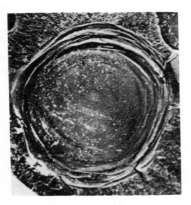

Fig. 1.7. Convoluted fecal string (arrowed) of a sediment-feeding metazoan. Upper Riphean, Ural Mountains. From Sabrodin (1971). Scale bar, 1 mm.

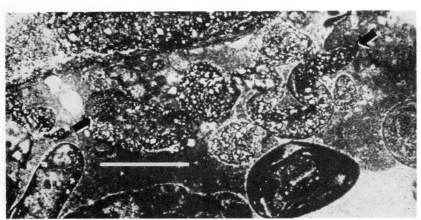

probably in the Jurassic, the diatoms in the Cretaceous, the dinoflagellates possibly in the Silurian or Permian and definitely in the Triassic. This makes the identification of Precambrian microfossils as members of extant taxa of Protista of the marine plankton impossible, in contrast to some extant Cyanophyta which can be recognized as belonging to it, with considerable probability, in Precambrian time. On the other hand, large numbers of organic walled microfossils are known which are placed in the 'Group Acritarcha' (Evitt 1963). This name indicates only that their affinities are unknown. Precambrian acritarchs are rarely older than Late Proterozoic. They are mostly simple spherical bodies with thin, smooth, organic envelopes. They are classed in a subgroup Sphaeromorphitae (Downie, Evitt & Sarjeant 1963). Their biological functions are unclear. Often assigned to this group, the genus *Chuaria* Walcott, 1899 has a remarkable geographic range throughout North America, northern Europe, the USSR, Iran, China and India. It occurs generally in considerable abundance in certain horizons of the Late Proterozoic and Vendian (Ford & Breed 1973, Hofmann 1977, Hofmann & Aitken 1979, Vidal 1981). It has a remarkable size range, from 0.5 mm to 5 mm diameter, according to Ford & Breed. A closely similar Late Proterozoic (mid-Vendian) fossil *Beltanelloides* Sokolov (Fig. 1.6) measures from 5 to 45 mm (Sokolov 1974). Its size makes it even more difficult to understand its growth and function as a phytoplankton organism. Sokolov considered it originally to be a 'medusa' but as it is a simple, originally spheroidal or more likely discoidal envelope without openings or appendages it is certainly not a coelenterate nor any other kind of animal, as Sokolov came to note later.

The earliest fossils which could be considered as animal remains (see review in Glaessner 1983) are unfortunately poorly known and incompletely described. Microfossils resembling Chitinozoa which may represent Protozoa or Metazoa have been mentioned (p. 20). Other microfossils occurring in the Lakhandin 'Series' of Siberia (950–1000 m.y. old) appear to be small metazoan organic structures. They were extracted by B. Timofeev and T. N. German from laminae of organic matter. They are still undescribed. All other metazoan fossils of that age are traces of organic activities rather than body fossils. A sigmoidal string of what appear to be fecal pellets about 0·37 mm in diameter (Fig. 1.7) from the middle Zilmerdak 'Series' (lower part of Upper Riphean of the Ural Mountains) was figured by Sabrodin (1971, 1972), together with a possible burrow. Clemmey (1976) described what he believes to be burrows from the Middle Roan formation of Zambia (about 1000 m.y. old) but Cloud (1978a; Cloud, Gustafson & Watson 1980) considers them as Recent termite burrows. Other questionable traces of bioturbation were recorded by Squire

(1973) from the Brioverian of the Channel Islands, considered to be about 750 m.y. old. Beer (1919) described a spiral measuring $13 \times 11$ mm preserved in relief on the lower surface of a slab of Rhotas Limestone of the uppermost Vindhyan of northern India which should be re-examined as it appears to be a trace fossil of metazoan origin. However, a Cambrian age of the Upper Vindhyan is possible. Supposed trace fossils from the Belt Supergroup in Montana (USA) have been redescribed as megascopic algae, aged 1300 m.y. (Walter, Oehler & Oehler 1976). A possible trace fossil from the Lower or Middle Riphean equivalents in the Ukraine, from a formation resting unconformably on a weathering horizon on volcanics which are 1000–1400 m.y. old, was described as *Rugoinfractus ovruchensis* by Palij (1974). Durham (1978, p. 37) considered it to 'merit special attention'. It is, however, also comparable with infillings of desiccation (syneresis) cracks which have been described repeatedly from Precambrian and younger rocks. The interpretation of a presumed medusoid fossil *Brooksella canyonensis* Bassler was questioned by Cloud (1968). Its subsequent assignment to the trace fossils (Glaessner 1969*a*) on the basis of its resemblance to *Asterosoma* was made more likely by Cloud's discovery of a second specimen. This interpretation as a possible 'infaunal deposit feeder' (Brasier 1979) was apparently confirmed by Kauffman (in Kauffman & Steidtmann 1981, p. 925) who gives the age as 1100–1300 m.y. although it is often stated as 900–1000 m.y.

The record of metazoan life from Early and Middle (?) Proterozoic rocks is apparently non-existent and that from 1000 to about 680 m.y. ago (Upper Riphean and equivalents) exists but is extremely poorly documented. The coincidence of the absence of metazoan remains in the early Middle Proterozoic and of definite eukaryotes favours the view that eukaryotes originated in mid-Proterozoic time or about the beginning of the Late Proterozoic, not in Early Proterozoic nor in Vendian time. We cannot yet assert as a fact that deposit feeders existed about 1000 m.y. ago but there are indications which make it more likely than the assertion of their absence. A fuller understanding of the distribution of structured fecal pellets in sediments should reveal the presence of sediment feeders more clearly than bioturbation, which may require more efficient structural adaptations of its originators to enable them to leave recognizable traces in ancient sediments. On the other hand it is probable that the earliest Metazoa were small detritus feeders. They may have developed cuticles of structural proteins which could be preserved together with other organic debris and discovered by palynological methods (acid treatment and bleaching) in unmetamorphosed sediments.

*The Vendian faunas.* The latest Precambrian (Vendian, about 680–560 m.y. ago) presents a clear picture of distinctive palaeontological documentation of a stage of metazoan evolution. Metazoans are widespread and locally abundant in rocks of this age (Chapter 2). Extended and thorough investigations have shown that a considerable number of sequences of Precambrian strata of that age contain distinctive fossils. Most of them became known only in the last 20 or 30 years. They have very little in common with the succeeding Cambrian faunas. The most noteworthy facts about the Metazoa of the uppermost Precambrian are the general absence of mineralized tissues such as calcareous or siliceous shells and the presence of representatives of several phyla, most of them extant in the living fauna. These are the Cnidaria, Annelida, Echiura, Arthropoda, Pogonophora (?) and possibly one or two others (Glaessner 1979*a*,*b*). The absence or probable absence of some phyla which are well represented in the Early Cambrian fauna, such as Porifera, Archaeocyatha, Mollusca, Brachiopoda and Echinodermata, and of the Platyhelminthes which on theoretical grounds are believed to have existed in Late Precambrian time, is noteworthy. The absence of Bryozoa and Chordata which make their first appearance in later Cambrian time should also be noted.

Late Precambrian metazoan body fossils are at present known from about 20 different regions in Africa, Asia, Australia, Europe and North America (Fig. 1.8). In some of these regions (southwestern Africa, South Australia, northern Russia, Ukraine) there are numerous localities where such fossils have been found, while from others we know so far only single finds. At the present time some of these fossils are regrettably still undescribed and some descriptions are still unpublished. Most areas have some distinctive kinds of Precambrian fossils in common but from South Carolina we know at present of an occurrence of a worm-like organism not found elsewhere (Cloud, Wright & Glover 1976). It is remarkable that even a few species are common to areas which are now as far distant geographically as South Australia, southwest Africa and northern Russia. A possible explanation for this fact will be discussed in Chapter 3. However, there are marked regional differences and we are not dealing with a cosmopolitan fauna in a biogeographic sense. There is evidence of considerable palaeoecological differences between fossil assemblages. What they have in common, however, is their distinctive level of diversification, with possible indications of increasing diversity within the time span of the fossiliferous Late Precambrian strata at the different localities. To give an order of magnitude to the possible time span, we anticipate here

Fig. 1.8. Worldwide occurrences of Ediacarian fossil Metazoa. *Australia*: 1 – Ediacara, Flinders Ranges; 2 – Punkerri Hills, Officer Basin; 3 – Deep Well, southeast of Alice Springs; 4 – Laura Creek, southwest of Alice Springs; 5 – Mt Skinner; 6 – Jervois Ranges, southwest Georgina Basin. *Africa*: 7 – southwest Africa/Namibia. *South America*: 8 – Mato Grosso, southwest Brazil. USSR: 9 – northern Russia; 10 – southwest Ukraine (Podolia); 11 – western Urals; 12 – Yenisey River (Igarka-Turukhansk); 13 – Irkutsk (Lake Baikal); 14 – Anabar and Olenek (northern Siberia); 15 – River Maya. *China*: 16 – northeast China (Longshan); 17 – Yangtze Gorge. *Northern Europe*: 18 – Lake Torneträsk (Sweden); 19 – English Midlands (Leicester). *North America*: 20 – southeast Newfoundland; 21 – North Carolina; 22 – northwest Canada (Mackenzie Mountains); 23 – central Iran. (Several localities without body fossils were omitted. Sequence of numbers follows text. Material from all localities except 8, 13, 15, 18, 22 has been available to me for study.)

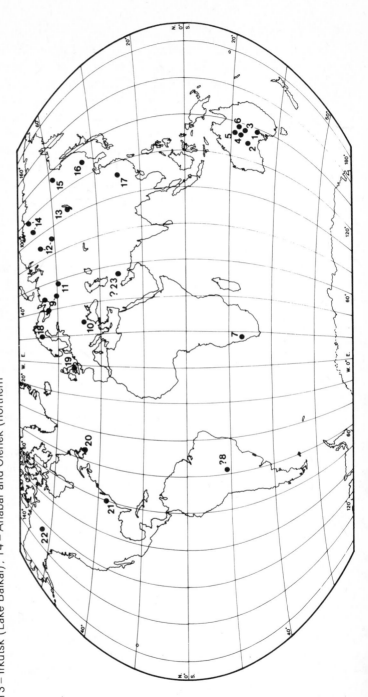

the results of an analysis of the datings of individual finds which will be discussed later. All these finds represent a time range of no more than 100 m.y., from about 580 to 680 m.y. before the present. It may be shorter but it is unlikely to be longer. It is no more than one fifth of subsequent Phanerozoic time, equal to the time which elapsed since the mid-Cretaceous. The fossiliferous rocks from this interval of time represent a variety of marine biotopes, but some of those which exist now, particularly those created by later differentiation of the biota, e.g. reefs, are absent, as are dark shales like the uncommonly richly fossiliferous Middle Cambrian Burgess Shale. The absence of most of the predators occupying the higher trophic levels of the modern marine faunas is of fundamental significance for the evolutionary level reached in Late Precambrian time. The definition of Metazoa as multicellular heterotrophs feeding by ingestion presupposes the presence of primary producers, presumably planktonic Protista, of which we have little direct knowledge. The abundance of life which can be documented at least at some sites must have led to an abundance of organic detritus and in turn to an abundance of detritus feeders, some of which can be documented by trace fossils of Late Precambrian age.

Between the very fragmentary record of traces of metazoan life about 900–1000 m.y. ago and the relatively abundant record of the latest Precambrian (680–580 m.y. ago) there is a significant time gap. It may be partly filled when more attention is given to a detailed search for traces of animal life activities in sediments 700–900 m.y. old. Pending further discoveries, we have to consider the possible effects of environmental changes during that time. In this connection we note that it is the time of the first of a number of Late Precambrian glaciations.

The sedimentary record of mid-Proterozoic time (1700–1000 m.y. ago) does not show any significant deviation from that of later times. It gives no grounds for assuming that the evolutionary origin of Eukaryota and the subsequent emergence of animal life, either as single-celled Protista or later as Metazoa, was due to some fundamental change in the environment. Once the oxygen level permitting oxidative respiration was reached, the adaptation to this efficient energy source was controlled not so much by changes of the external environment as by advances in the biochemistry and physiology of the organisms. We can speak of a phase of cytological evolution without contradicting any basic assumptions of modern selectionist (neo-Darwinian) evolutionary theory. Selection would favour physiologically more efficient organisms functioning in an unchanging environment. The significance of cytological changes in cell division leading to meiosis and redistribution of hereditary material by sexual processes in early eukaryotes has been discussed (Schopf *et al.* 1973,

Stanley 1976*b*). While palaeontological evidence is not as unambiguous as was once thought, it is clear that these steps in cytological evolution were determined by selective processes favouring more efficient reproduction and increased genetic variability in populations, without changes in their environment.

The picture changes when we approach the latest Proterozoic. Worldwide glaciations of Late Riphean and Early Vendian age (Chumakov 1978, 1981) are dated, at various localities, from not less than about 800 to about 650 m.y. before the present. Abundant evidence of metazoan body fossils is found in sediments which are *younger* than those containing evidence of the latest glaciation, in North America, northern Europe, China, southwestern Africa (where there may have been later cold periods) and Australia (Glaessner 1977) (see p. 102). This observation is not only of stratigraphic significance but is also likely to reflect influences on the history of the biosphere. The existence of rare traces of animal activities locomotion and feeding) and of still unidentified microscopic fossils of probable animal origin recorded from preglacial Late Proterozoic (Upper Riphean) sediments 1000–800 m.y. old supports the view that the Ediacarian fauna does not represent the earliest Metazoa. Even without these data, this is obvious from the relatively high level of metazoan differentiation. A considerable interval of time was required for metazoan 'prehistory'.

### 1.6 Origins and early differentiation of the Metazoa

Some zoologists have expressed the view that the study of fossils is useless for investigation of phylogeny because of the incompleteness of the fossil record. They may have overlooked the fact that zoological knowledge of the great majority of a million living species is incomplete in respect of vital facts of their genetics, molecular biology, ethology and even morphology. That does not necessarily limit the value of their research or make it essentially speculative. The existence of certain kinds of animals, whether expected or unexpected, whether well or incompletely known, at certain times in the past must be taken into consideration. Without historical background, without palaeobiology, the pursuit of biological science must remain incomplete. The study of the fine structure of mineralized tissues of living organisms was aided and stimulated by palaeontological work, and the systematics of many major groups of animals was significantly modified and refined by the integration of palaeo- and neobiological knowledge. What has been discussed so far in this chapter was based on integration of theories and observations in many fields with a fossil record that was until recently thought to be non-existent.

It was shown to exist although it is less complete the further back we go into the past. It is appropriate to speak of a prehistory of the biosphere, as we speak of the prehistory of mankind when we go back beyond the periods for which we have written documentary records. In order to understand the earliest preserved abundant assemblages of Precambrian Metazoa in relation to the history of life at earlier Precambrian times, we shall briefly consider some of the many biological theories of origins of Metazoa and their early differentiation. These theories are essentially phylogenetic since we are trying to trace the course of evolution, the evolutionary pathways from one group or groups, extinct or still living, to others. Phylogenetic theories and the theory of phylogeny on a supraspecific level are once again becoming respectable subjects of scientific enquiry and debate (Dougherty 1963, Brien 1969, Eldredge & Gould 1972, Jägersten 1972, Gould 1977, Hallam 1977, Hanson 1977, Heberer 1967–1971, Riedl 1978, Stanley 1979, and many others), but this is not the place for a critical review of evolutionary or phylogenetic theory. The varieties of applied methodologies and of the results presented, often in the graphic form of taxa linked by presumed lines of descent, is bewildering. Approaches and suggestions which appear to be less controversial than others are here considered with the specific purpose of providing a framework for the linking of prehistorical and historical data on the early evolution and differentiation of animals. The outline of our knowledge of Precambrian fossils presented above makes it clear that phylogenetic discussions relevant to them must be kept on the level of higher taxa. No significant evidence for evolution at the species level in Precambrian time has been discovered. Omission of discussion at this level does not necessarily imply reliance on special mechanisms of 'macroevolution' as distinct from evolution of species.

Hanson (1977) has laid down three 'procedural guidelines' for phylogenetic analysis. Taking for granted, without questioning it, because it is considered as outside the scope of our discussion, the first rule 'Species define the basis for interorganismic comparisons', we turn to the second one. It says: 'Phylogenetic comparisons must examine the exploitive, homeostatic, and reproductive functions and/or their anatomical correlates whenever possible'. The third rule states that 'Every innovative step must be selectively advantageous'. These guidelines (a better term than rules) are necessary constraints which can lift phylogenetic investigations above the level of speculation. They are here rephrased in order to make them self-explanatory. Hanson's third rule means that every innovative step implied in the proposed phylogeny should be seen as a possible result of a selective pressure. Rule two means that phylogenetic analysis is based on functional morphology, the primary functions being (*a*) bioenergetically and environmentally conditioned, (*b*) reproductively advantageous in their result and (*c*) constrained by homeostatic principles. Homeostatic in this context means preserving constancy of internal environment generally, and of conditions for development and growth specifically, during evolutionary changes in the form–function complex. The significance of the basic functions of the animals and their expression in morphology are discussed in most textbooks of biology and need no further explanation here. The homeostatic principles guiding phylogeny concern, firstly, the existence of homologous structures in different organisms. The concept of homology cannot be put aside as allegedly based on circular reasoning. It is an essential operational tool of phylogenetic analysis (Boyden 1947, Remane 1956, Simpson 1961, Peters 1972, Hanson 1977, Riedl 1978, Rieppel 1980). Secondly, homeostatic principles underlie the channelling of development discussed extensively by Waddington in many of his works, and ultimately the relations between ontogeny and phylogeny (de Beer 1951, Gould 1977). Again, this has been brushed aside mistakenly, as based on the supposedly long-discredited biogenetic 'law' of Haeckel. Thirdly, they underlie the morphotype or better groundplan or structural plan of organisms, a concept which is more often invoked in German than in Anglo-Saxon literature. It has no necessary epistemological connection with idealistic morphology but can be based on genetics and information theory as well as on considerations of biophysics and materials science as applied to organic form and function (Pantin 1951, Gould 1970, Peters & Gutmann 1971). The necessity for every part of an organism to function in accordance with structural and mechanical principles, within the constraints of strength of available materials, and to preserve ability to function throughout all stages of growth and in response to all tolerable environmental stresses and the exigencies of competition is generally admitted. All this can be expressed briefly as structural plan ('Bauplan') or its synonyms but care must be taken to avoid the literal, typological, theological or teleological implications of the words 'plan' or 'type'. When reference to engineering principles is made, it should be remembered that engineers order materials according to specifications, and mathematical analysis precedes design. Evolution proceeds with inherited material, by trial and error. It has often been said that it is 'opportunistic' in the choice of environments for its products. Careful consideration of functional plans in the sense here indicated is particularly important when the deformable soft bodies of extinct Precambrian Metazoa have to be interpreted as formerly functioning organisms.

The gap in fossil evidence between Metazoa and their presumed protozoan ancestors can be filled at the present time only by biological theory. Hanson (1977) is mainly concerned with the interrelations of the Protozoa and their possible origins from other Protista, most of which remain unidentified. Concerning metazoan origins, he derives the Porifera from choanoflagellates, which is a widely accepted view, and the turbellarian platyhelminths from primitive ciliates, considered as descended from unidentified zooflagellates. The Cnidaria are questionably derived from amoeboid zooflagellates. The very large amount of factual data from the living Protozoa and lower Metazoa contained in Hanson's voluminous and thoughtful work cannot be reviewed here. While it throws as much light on the world of animals at dawn as can be derived from present knowledge of their living representatives, it stresses repeatedly, as various groups are reviewed, the scantiness or total absence of their palaeontological record. On present knowledge and with present methods, the distinction between Protozoa and other Protista on which Hanson insists (for other views on systematics see Whittaker & Margulis 1978) cannot be made for those microfossils which have no living representatives. Precambrian microfossils not assignable to blue-green algae, eukaryotic algae and fungi may have no living representatives. However, considered views on the possible origin of Metazoa from Protozoa can be used as a guide to further search which may reveal modes of fossilization as unexpected as that of the rich Gunflint assemblage of Prokaryota.

Comparative embryological studies of phylogeny of the lower invertebrates were critically reviewed by Ivanov (1968). His short but important book is concerned with the origin of the multicellular animals. A critical history of phylogenetic theories is followed by an exposition of the author's own views (Fig. 1.9). They evolved from the work of the famous Russian school of embryologists of the last century (Kovalevsky, Metchnikoff and others) and their modern successors, including the comparative morphologist Beklemishev (1964). Work in other countries, from Haeckel to Hanson, up to the 1960s, is given due attention. Ivanov's proposed stages in the phylogeny of the Metazoa can be considered as a model bridging the gap in metazoan prehistory to which reference was made above (p. 30). It can be tested, to some extent, at least against biochronological and palaeobiological facts, if not against palaeomorphological data. One of the reasons for the absence of such data becomes obvious when the (unstated) scale of size of the organisms shown in Fig. 1.9 is considered. In dealing with the problem of the derivation of small multicellular from single-celled organisms, Ivanov argues convincingly against theories of cellularization (i.e. the separation within multinucleate protistan cells of compartments destined to become cells of Metazoa). This course of evolution is still

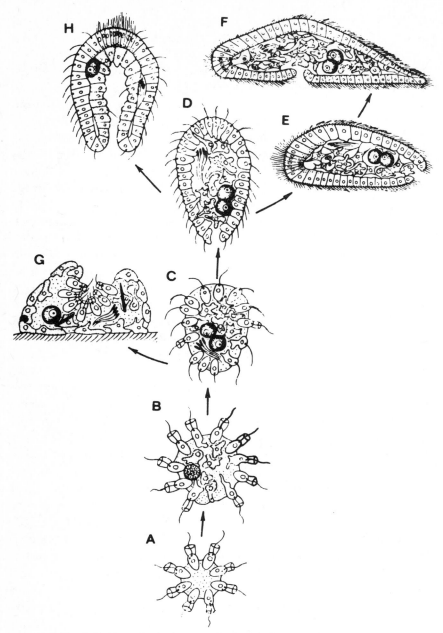

Fig. 1.9. Hypothetical stages in the phylogeny of the lower Metazoa. (After Ivanov 1968, Fig. 118; 1970, Fig. 8.) A, Colonial choanoflagellate (*Sphaeroeca*-type); B, the same, *Proterospongia*-type; C, early Phagocytella; D, later Phagocytella; E, ancestral benthic turbellarian flatworm with mouth and bilateral

accepted by Hanson. Ivanov starts his phylogenetic construction with a free-swimming spherical colony of undifferentiated animal flagellates, resembling the living *Sphaeroeca* belonging to the class Craspedomonadida (or Choanoflagellida). Reproduction was asexual, by separation and division of individual cells and formation of new colonies. This stage in evolution is followed by a colonial stage with increased integration of the cells. Somatic and reproductive cells are differentiated, sexual reproduction leads through cleavage of the zygote to the formation of a blastula-like organism with multiplication of somatic cells by mitotic division and development of radial symmetry. *Proterospongia haeckeli* Kent is said to show some approximation to this stage. It is a rare organism assigned to the choanoflagellates and considered to be close to the ancestry of Porifera. It requires further study (Ivanov 1968, p. 223, Hanson 1977, p. 452), particularly concerning its reproduction. The next stage is frankly hypothetical as an adult form. It is based on Metchnikoff's Phagocytella which he had earlier named Parenchymella and which is similar to the planuloid stage in the ontogeny of various lower Metazoa. In Metchnikoff's and Hyman's (1940) view it is considered as a free-swimming organism, a possible ancestor of the Eumetazoa. During its embryogeny a blastula-like larva was supposedly formed which grew through cell division. Differentiation of ecto- and endoderm occurred through multipolar immigration of specialized cells into the interior of the organism. Among them were phagocytes moving in a parenchyma. Thus there is at that stage a differentiation into an external, locomotory *kinetoblast* and an internal, digestive *phagocytoblast*. It is noteworthy that Ivanov accepts the notion of a solid Phagocytella originally without a mouth while objecting firmly to Haeckel's generalized gastraea theory and describing Lankester's assumption of a digestive cavity in a 'Planula' without a mouth as 'obvious nonsense' (Ivanov 1968, p. 133). However, according to Ivanov a mouth is necessarily formed subsequently in this Phagocytella at the posterior end of this free-swimming organism where food particles in the water driven by the flagella of the kinetoblast congregate in the 'backwash'. It follows from this view that the Porifera originated prior to this stage, becoming totally sessile, with the kinetoblast transferring its hydrokinetic

Caption for Fig. 1.9 (*cont.*)
symmetry; F, ancestral acoeloid turbellarian, with enhanced cell differentiation and ventral mouth. Sidelines from stage B leading to G, a primitive sponge, sessile, locomotion replaced by hydrokinetic function; and from stage D leading to H, a primitive coelenterate (Gastraea). (No scale, note small number of cells in each section, hence microscopic size of organisms.)

function into interior cavities of the organism. The Turbellaria Acoela are also close to the 'Phagocytella' but remained motile by ciliary locomotion and originated after it had evolved a mouth. Their internal organization, in the absence of an intestine, corresponds according to Ivanov (1968, p. 263) to the phagocytoblast of Metchnikoff's Phagocytella. This hypothetical but potentially functioning organism also gave rise to the swimming Ctenophora and, through adaptation to a sessile mode of life, to the Cnidaria. Ivanov considers single hydroid polyps as their ancestral forms. All 'coelenterates' have acquired diploblastic organization. Only their ancestral forms are described by Ivanov as 'gastraea-like'.

Jägersten (1972) expanded his earlier 'Bilaterogastraea' theory into a complete phylogeny of the Metazoa based on an untestable hypothesis. He states that two phases of their life cycle arose when 'the adult of the primeval ancestor of the metazoans, viz. the holopelagic, radially symmetrical Blastaea, descended to life on the bottom' and became bilateral as a Bilaterogastraea with coelomic pouches derived from gastric pouches of coelenterates. The 'juvenile stage remained in the pelagic zone' (p. 216) in most metazoans, not by adaptation to dispersal but because of recapitulation of the postulated primitive ontogeny. Jägersten's simplistic and formalistic theory on which Hyman (1959, p. 759) made cogent critical remarks but which has found some supporters, implies incipient formation of a coelom and metamerism in coelenterates. This was unacceptable to Clark (1964, 1969, 1979) whose views are based mainly on functional considerations, relating primarily to locomotion as a distinctive basic problem in metazoan diversification. Clark's position on the earliest history of the Metazoa is somewhat agnostic (1979, p. 63) as was Hyman's earlier conclusion (1959). Hanson (1977) does not accept the Turbellaria Acoela as direct ancestors of all other Metazoa including the coelenterates, considering with the great majority of zoologists Hadži's (1963) theory as 'discredited', but he places a question mark at the origin of the Cnidaria, where others (Hand 1959, Hyman 1959, Ivanov 1968) had placed a planuloid organism. Clark (1979) concerns himself with the radiations of the acoelomates and coelomates which we shall discuss (in Chapter 3) after a review of palaeontological data and their chronological implications (Chapter 2). However, several theoretical viewpoints must be briefly considered before proceeding to these matters.

(1) Phylogenetic theories. A generalized historical background to the still controversial questions of invertebrate phylogeny was given by Clark (1964) who later reviewed (1979), in the light of his studies on coelom formation and metamerism, arguments and constructions presented in the course of lively debates in the 1950s and 1960s from the viewpoints of

comparative functional morphology and embryology (or developmental anatomy). This was followed by a fierce debate, mainly in the German literature, between classical morphologists (Reisinger 1970, Siewing 1972, Ulrich 1972, Remane 1973) and the 'Frankfurt School' (see Gutmann 1977, 1981, and references therein) whose views are based dominantly and almost exclusively on functional–biomechanical analysis and model construction. This continuing debate will not be reviewed here (although it is significant and mostly constructive) because it was so far largely concerned with the origin of the chordates. This is essentially outside the scope of our examination of the Precambrian–Cambrian evolution of the Metazoa.

Phylogenetic hypotheses concerning the transition from Protozoa to early Metazoa have advanced from early speculations or subsequently refuted observations to more sophisticated models proposed in recent years. Combined evidence from histology, embryology, comparative and functional morphology, physiology and ecology of living organisms is used extensively. *The construction of hypothetical ancestral organisms on a purely morphological or embryological basis without full exploration and explanation of their functioning* 'ab initio' *and of the selection pressures leading to their origination and transformation into descendant taxa is being discouraged.*

(2) Palaeoecological considerations. There is at present hardly any evidence for the nature of the Late Precambrian metazoans prior to the radiations which the Ediacarian faunas reflect and which will be discussed later (Chapter 2). There are, however, hypotheses about how and where the early metazoans lived. Palaeoecological theorizing based on living taxa of the lower Metazoa is notoriously hazardous. Their survival may be due to their limitation to environments hostile to predators when the latter appeared later. An example among Prokaryota is the 'survival' of columnar stromatolites in hypersaline environments which tells us much about stromatolites and their cyanophyte builders but little about the habitat of their Precambrian precursors. This example does not necessarily lead to a refutation of the hypothesis that some of the adaptive radiations of the lower Metazoa occurred in the reducing environments existing within the substratum of marine sediments. The planuloid hypothesis concerns the origination of the early Metazoa but does not explain their differentiation. Some of them would have contributed to the zooplankton while others could have descended to the sea floor and developed the ciliary creeping habits of Platyhelminthes. The organic remains of microplankton are incorporated in the sediments. In the absence of mixing with oxygenated water, or when the oxygen is exhausted by their decomposition, an anoxic environment is produced below the sediment/water interface. This is now

known to be occupied by a varied biota (Fenchel & Riedl 1970, Rhoads & Morse 1971, Boaden 1975, Clark 1979). This infaunal biota (Fig. 1.10) includes Platyhelminthes (Turbellaria Acoela and Catenulida) and Gnathostomulida in the sulfide biome or thiobios of near-shore and marginal marine zones. Representatives of additional, more advanced taxa occur when and where more oxygenated sediments are deposited above

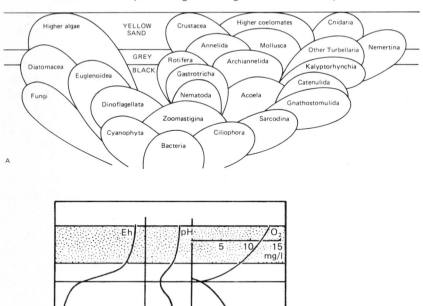

Fig. 1.10. Depth distribution of benthic marine taxa in relation to deoxygenation of sediments. A: The names are plotted according to maximum depth of occurrence relative to the 'redox potential discontinuity layer', here labelled 'grey'. Lines around names suggest 'relationship between depth and evolutionary origin' according to the author of the figure (Boaden 1975). B: Schematic representation of Eh and pH profiles in sediment from near-shore and restricted marine environments. Part of Fig. 4 of Fenchel & Riedl (1970). The two horizontal lines across the middle of the figure delimit the 'grey' zone of A. (Rhoads (1974) has reversed the terms 'black' and 'grey'. Note that these sediment layers do not normally represent rock sequences but presently observed stages of syndiagenesis of sediments. Under unchanging conditions the zones and their biota will move upward with continuing sedimentation, producing dark, organic-rich clastics.)

reducing black muds. Considering the subsequent colonization of the hostile sulfide-enriched environment by secondarily and convergently adapted members of other phyla as a subsequent modification of the biota, Fenchel & Riedl, supported by Boaden, conclude that the occurrence there of many primitive Metazoa is a relict of a biosystem which preceded the fully aerobic biosphere. The probable attainment of an atmospheric level of at least 1% PAL of oxygen about the time of the origin of the Eukaryota makes it feasible to link the acoelomate radiation with infaunal habitat and to consider some of the present sulfide biota as hypothetical relicts of Late Precambrian (pre-Vendian) faunas. The absence of Cnidaria from such environments indicates not a later origination of coelenterates but other, concurrently occurring adaptive radiations in co-existent, more oxygenic environments. At any time during the development of the oxygen content above 1% PAL there were opportunities for the capture of nutrients from primary producers, i.e. the phytoplankton anywhere in the water column from the surface to the bottom. Sediments formed at the relevant time indicate the existence of normal conditions of oxygenation of sea bottom water: black shales which formed under reducing conditions are not prevalent among the terrigenous formations of the Upper Proterozoic, hence the sulfide biome was not an exclusive or particularly likely environment for evolutionary developments. An hypothesis requiring this lacks geological or palaeoecological verification.

Summarizing the present evidence from biological observations for metazoan prehistory we reach the following result. Boaden's (1975) conclusion that the Gnathostomulida, Platyhelminthes and Aschelminthes originally evolved under dysaerobic conditions can be accepted, assuming that dysaerobic means atmospheric oxygen levels of a few percent of the present level. Boaden's further conclusion that the early Metazoa were benthic anaerobes does not necessarily follow from either his own or geological evidence. The rate of oxygen accession to the atmosphere and to bottom waters and the distribution of oxygenation in benthic habitats at the relevant time are not sufficiently well known to support conclusions based on observations in the present near-shore thiobiome alone. There is no need to confine early metazoan evolution to this type of habitat. The later origin of the coelenterates compared with that of the Bilateria is also unacceptable. The basically sessile benthic coelenterates were not the likely ancestors of the locomotory, bilateral metazoans. It should be remembered that for good reasons Hyman (1959, pp. 750 ff) considered the derivation of the coelom from the coelenteric pockets of the Anthozoa (rugose corals), which was updated by Remane from nineteenth-century notions and accepted by some of his followers, as 'fantastic nonsense' and Jägersten's bilaterogastraea theory as 'absurd'. For these reasons *Ivanov's hypothetical phylogeny of the earliest Metazoa is here considered as acceptable and Clark's discussion of subsequent radiations will be the basis of further comparisons with documented events of Late Precambrian time* (Chapter 3).

(3) Relative timing of metazoan evolution in the Late Proterozoic. A common assertion in many phylogenetic theories is the independent origin of the Porifera from choanoflagellates. Lacking organized tissues and possessing the potential of regenerating the entire organism from the smallest fragments, they are often separated from the Eumetazoa as belonging to a subkingdom Parazoa (Margulis 1974). The implication is that their origination would be an early event in the history of animals. The fact that most of them possess mineralized spicules would lead to the expectation that their remains would be found in Late Precambrian sediments. This expectation has not been fulfilled. One or two reports of Precambrian sponge spicules have not yet been disproved but they are questionable. Three explanations of these observations are possible. (i) The sponges could have evolved early with bodies strengthened by organic fibres without significant fossilization potential. Abundant living sponges possess only spongin fibres. (ii) The sponges with spicular skeletons could have evolved early but remained undetected in the fossil record because of small size and lack of distinctive characters. (iii) They could have evolved at the end of Precambrian time when the diversification of many animals with mineralized tissues took place. Assumption (iii) is supported by analogy with a similar late development in the Foraminifera and Radiolaria which evolved rapidly from few structurally primitive early Palaeozoic representatives. This hypothesis would not necessarily exclude (i) if in fact the Porifera without mineralized spicules can be considered as ancestral to spicule-bearing major taxa.

The second postulate is the independent early origination of the coelenterates which are also known as Radialia or as diploblastic Metazoa. No fossil Ctenophora are known and therefore only the phylum Cnidaria will be considered here. The high degree of taxonomic diversity attained by the cnidarians in the Ediacarian faunas (Chapter 2) indicates a long preceding period of existence. The Phanerozoic record of soft-bodied Cnidaria is remarkably patchy. Suitable conditions of preservation could have been similarly rare in Precambrian time. The evolutionary rate of diversification being unknown, the 'long' period of cryptic evolution is to be seen only as a measure of time elapsed in relation to the time span required for the diversification of other Late Precambrian Metazoa. Those definitely known include one representative of the phylum Echiura, a number of annelids and a few primitive arthropods. The annelids are

*Origins and early differentiation of Metazoa* 41

Fig. 1.11. Outline of metazoan phylogeny assuming acoeloid/planuloid origins and polyphyletic origin of the coelom. From Clark (1979, Fig. 9).

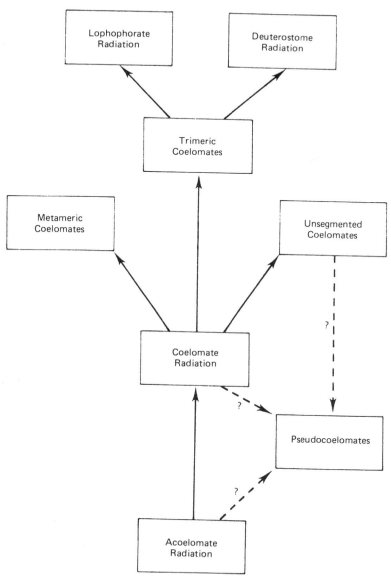

represented by at least six genera belonging to at least four very different families. None of them could possibly be considered as primitive by comparison with living members of this phylum. This contrasts sharply with the representation of the arthropods. Only two kinds are known, one representing primitive trilobitomorpha or chelicerates, the other primitive crustaceans. *It must be concluded that the origination of the arthropods was closer to the age of the Ediacarian faunas than that of the annelids which in turn was closer than that of the cnidarians.*

Most studies of the evolution of the living fauna make it likely that the diversification of acoelomate metazoans (mainly Platyhelminthes, Nemertina and Gnathostomulida) preceded that of the coelomates (Fig. 1.11), though Clark (1979, pp. 95–7) leaves the other alternative open. The acoelomates and pseudocoelomates ('Aschelminthes': Nematoda, Nematomorpha, etc.) are entirely or almost totally absent from the fossil record including that of the Precambrian. Omitting the living non-marine and parasitic forms from consideration, we note that these lower free-living Metazoa are small, many of them millimetres in size, and their tissues are extremely soft. These are sufficient reasons for their absence from most kinds of sedimentary rocks. Future lucky finds of some remains of some of these animals are not impossible but it cannot be concluded that they did not exist prior to the appearance of the coelomates because of their absence from the Precambrian record of life. Tubes of worm-like animals (Sabelliditida) may occur in pre-Vendian sediments (see p. 89). The Mollusca are believed to be non-metameric descendants of Platyhelminthes (Salvini-Plawen 1969, Clark 1979). No molluscan body fossils are known from the Precambrian but their locomotion traces could and probably do occur. *The presence of metameric coelomates in the Late Precambrian (Ediacarian) faunas is obvious* (Chapter 2).

# 2
# The Ediacarian faunal assemblages: discovery, composition, significance

## 2.1 Discoveries at Ediacara

In 1946 R. C. Sprigg, who was then the Assistant Government Geologist of South Australia, examined old lead mines in the Ediacara Hills, a desolate, low range in arid country some 600 km north of Adelaide. The main objective of his work was to review the possibilities of re-opening old silver–lead mines, in the course of a re-assessment of the state's mineral resources. Having acquired the habit of fossil collecting during his student days, Sprigg kept his eyes to the ground even when traversing unpromising quartzite outcrops a short distance southwest of the old mines. To his surprise he found numerous casts and impressions of 'jellyfishes'. He described some of them briefly in the following year, noting correctly that they are 'among the oldest direct records of animal life in the world' and that 'they all appear to lack hard parts and to represent animals of very varied affinities' (Sprigg 1947, p. 212). More fossils were collected soon afterwards by the discoverer and by Professor Sir Douglas Mawson of the University of Adelaide and his students, at the same locality and its northern extensions. They were described in some detail by Sprigg (1949). He stated (p. 73) that there can be no hesitation in placing many of the forms with either the Hydrozoa or Scyphozoa and added: 'Some forms are referable to algae but these will not be described in this paper'. Sprigg's interest in his remarkable finds continued to the present time but his extensive subsequent work in regional geological and mineral exploration and numerous other varied and fruitful activities did not leave him time for further palaeontological research.

What happened some 10 years later was recorded by Hans Mincham (1958), then a school teacher and later a staff member of the South Australian Museum and a well-known author and naturalist. 'It was just after Christmas in 1956 that Mr. Ben Flounders and I set out to visit...Ediacara.' Mr Flounders was a keen mineral collector and an able photographer. 'Unfortunately, we ran into a heat wave – the worst possible weather for fossil hunting at Ediacara. Because of the extreme heat it was impossible to hold any piece of rock picked up from the ground.' The collectors' first finds were encouraging and they returned at a more suitable time, in September 1957, when they made a large collection of fossils which they presented to the South Australian Museum. Some of Ben Flounders' excellent photographs were submitted to me for identification and I informed the finders that they had discovered entirely new kinds of animals that had never been seen before (see Glaessner 1958). In March 1958 the Museum sent out an expedition led by Dr Brian Daily, then Curator of Palaeontology. The six men of the expedition, including the two amateur collectors and one of my graduate students, collected in four days enough fossils to fill two small trucks and a trailer. In May 1958 the area was proclaimed a fossil reserve under the control of the State Minister of Education and the South Australian Museum. In October 1958 the first joint expedition of the museum and the Geology Department of the University of Adelaide was carried out (see Glaessner & Daily 1959) and systematic exploration at Ediacara continued for a number of years, resulting in the collection of over 1500 specimens of fossils. The area is now covered by the regional mapping carried out by the Geological Survey of South Australia (Leeson 1970, Forbes 1972, Coats 1973). As would be expected, fossils are no longer found by the truckload and collectors, who have to be authorized by the State Minister for the Environment, have to be content with a few specimens. However, other fossiliferous localities have been discovered in the Pound Quartzite which must have been deposited over a large area of the northern and central Flinders Ranges in South Australia (Fig. 2.1).

The Pound Quartzite (or Sandstone) is by definition the highest stratigraphic unit of the 'Adelaide System' which at the time of Sprigg's discoveries and up to their first revision (Glaessner & Daily 1959) was thought to be of Late Proterozoic to Early Cambrian age. It was named by Mawson (1938) who measured a section in or near the Parachilna Gorge and recorded the occurrence of worm tracks below the top of the section. He considered tentatively, in the absence of other evidence, that it should be included in the Cambrian as its lowest stratigraphic unit. Above it are limestones rich in Archaeocyatha, of Lower Cambrian age. The name was taken from the Wilpena Pound, a spectacular topographic feature 40–60 km to the south, where the quartzite forms the upper reaches of an elliptical mountain wall over 1000 m high around a syncline. At Ediacara the Pound Quartzite is overlain by a stratigraphic sequence which was believed to be

### Discoveries at Ediacara

Fig. 2.1. Map of fossil discoveries in the Flinders Ranges. Stippling indicates outcrop of Pound Subgroup rocks. Fossil localities indicated by F. After Jenkins *et al.* (1983).

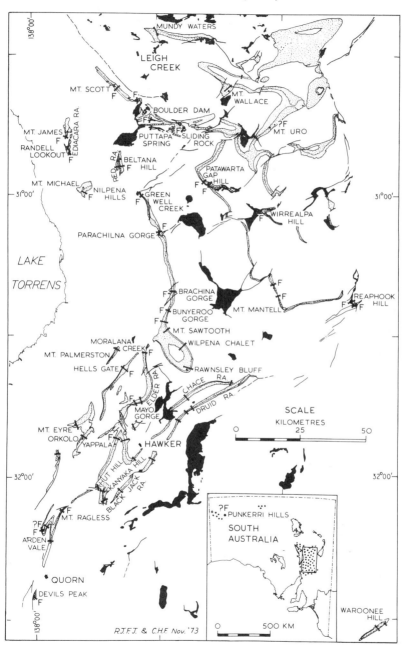

## 2: The Ediacarian faunal assemblages

transitional to the Archaeocyatha-bearing dolomites and limestones. Subsequent regional mapping by Dalgarno (1964) showed that this supposedly transitional contact above the Pound was actually disconformable. He named the unit the Parachilna Formation, which is characterized by the abundant occurrence of U-shaped burrows (*Diplocraterion*). Some of them penetrate the top of the Pound Quartzite, which proves that it must have been soft sand in Early Cambrian time. Mary Wade (1970) discovered

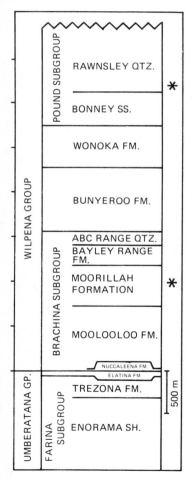

Fig. 2.2. Generalized stratigraphic column of Late Proterozoic in central Flinders Ranges of South Australia showing most of the Marinoan Series. Occurrences of metazoan fossils marked *. Upper asterisk marks position of Ediacara Member. (Modified from Jenkins *et al.* 1981. Each scale division – 500 m).

numerous occurrences of components of the Ediacara fauna at various localities in the Flinders Ranges and concluded that they were confined to a stratigraphically correlated horizon in the Pound Quartzite, up to 112 m thick, above the base of its upper, mainly white member which overlies a dominantly reddish member. These two divisions were formally named by Forbes (1971) the Bonney Sandstone overlain by the Rawnsley Quartzite (Fig. 2.2). Later workers tend to consider these divisions as formations of the Pound Subgroup of the Wilpena Group and the silty, fossiliferous unit as the Ediacara Member (Jenkins 1981). A further important development in the stratigraphy of the Cambrian rocks above the Pound Quartzite was the recognition by Daily (1973) of the Uratanna Formation of sandstones and siltstones with Cambrian trace fossils. This formation overlies the Pound with erosional unconformity. In places, erosion has removed most of the Rawnsley Quartzite down to the level of the Ediacara fossiliferous strata. The total thickness of the Pound Quartzite varies, partly because of original conditions of sedimentation and partly because of erosion of its top. Maximum thicknesses up to 3000 m have been observed (Coats 1973) but in most of the area its maximum is nearer 1000 m. The fossiliferous strata occur at Ediacara a short distance above the base of the Rawnsley Quartzite which is here about 300 m thick. Both the Uratanna and the Parachilna Formations are correlated by Daily with the Tommotian Stage at the base of the Cambrian (see Chapter 4).

## 2.2 Palaeoenvironment and fossilization

The extensive mapping and stratigraphic work during the last 20 years has not only fully confirmed the relative dating of the Ediacara fossil assemblage; it has also confirmed the original conclusions concerning the conditions in which this remarkable fauna lived, died and was preserved in the rocks. Sprigg (1947, p. 214) concluded from his observations that 'The enclosing sediment was originally a fine-grained and well-sorted sand which had accumulated near the western margin of the vast Flinders sedimentary geosyncline. The environment of entombment was that of intertidal flats or of the strandline.' It should be noted that in modern terms most of the sand would be described as medium-grained. Sprigg noted correctly that all fossils were found on 'free' faces of flaggy quartzite, i.e. those freed by weathering, but he believed erroneously that they were 'always on the upper surfaces of these slabs'. This impression was corrected by Mincham (1958) and explained by Glaessner (Glaessner & Daily 1959). Sprigg (1947, p. 215) came to the following conclusion: 'The fissility of the quartzite appears to have been controlled to some extent by

## 48    2: *The Ediacarian faunal assemblages*

the presence of clayey films, which might also have had significance in the preservation of the fossil impressions'. The alternation of thin clay layers and thicker sand lenses (Fig. 2.3) constitutes the 'very special conditions of burial' which Sprigg and some later commentators held responsible for the presence of fossil remains of the former existence of 'jellyfishes and related soft-bodied coelenterates'. Their abundant occurrence at Ediacara seemed remarkable, particularly in view of the fact that living medusae consist of 96–98% of water (Hyman 1959, p. 722). Their mesogloea contains collagen fibres which are considered by Hyman to be comparable to the connective tissue of higher animals. The body of a medusa can be so tough that I was able to place my entire weight (80 kg) on a recently stranded specimen, about 20–25 cm in diameter, on a beach. This resulted in no damage to the medusa but produced a clear impression of its oral surface on the sand. A medusa preserved on the lower surface of a quartzite slab, in positive hyporelief according to a terminology developed for trace fossils, is in fact a sand cast or mould infilling a depression in a substratum of clay or silt which was caused entirely or mainly (in the case of composite moulds) by its exumbrellar surface. An impression of a body fossil in

Fig. 2.3. Typical exposure of fossiliferous Ediacara Member of Rawnsley Quartzite, Ediacara Hills. Flaggy, ripple-marked, cross-laminated sandstones and quartzites.

negative hyporelief must be caused by the body still resting on (or stuck to) the substratum when it was covered by a shifting sand wave. The fossiliferous phase or member of the Rawnsley Quartzite consists of interbedded and cross-bedded flaggy sandstones with silty partings, or of sandstones and siltstones, while the rest of the Pound succession consists of silty sandstones, massive sandstones and quartzites. The fossiliferous sequence represents an interval of deposition under conditions of temporary reduction of hydrodynamic energy in the area. Silt and clay could settle in limited pools or lagoons between sand ridges which were frequently moved on by currents, burying the pools and the remains of animals and traces of their activities. The weight of several hundred metres of sand eventually accumulating above them compacted and altered the silt and clay lenses to such an extent that no animal remains could be preserved in them. They remained in the sand mainly as casts and moulds. It acted like a foundry sand from which moulds for the casting of metals are made in the metal industry. The composition of foundry sand must conform to exact specifications as to grain size and shape and clay content. The sedimentology of the Pound Quartzite has not yet been studied in sufficient detail to specify the exact condition for the casting of soft-bodied animals preserved during the relatively quiescent time interval represented by the Ediacara Member. It is not surprising that its fossil content varies from place to place quantitatively and qualitatively. The greatest abundance is found near the western margin of the Flinders Ranges, presumed from palaeogeographic considerations to be closest to the shore of the basin of deposition. This was Sprigg's 'Flinders Geosyncline' which is now more generally known as the Adelaide Miogeosyncline. There were intermittently rising and sinking zones within it. If there was a shoreline near Ediacara, which is likely because evidence of local emergence above sea level is found in the form of rare drying cracks, it was not the only one in the region. The water during the time of deposition of fossiliferous strata was shallow. Goldring & Curnow (1967) thought it was not less than 6 m while Jenkins, Ford & Gehling (1983) concluded that the fauna was deposited mainly in the intertidal zone. Because of the wide areal spread of similar faunas, with planktonic, nektobenthic and sessile benthic components in almost complete growth series indicating near-autochthonous habitats, it seems inappropriate to specify the local depth of water too closely. Silty pools alternated with sand waves and the site of the shoreline could have moved during the deposition of the fossiliferous strata. Swarms of medusae drifting shoreward alternated with groves of sessile pennatulid-like organisms; populations of nektobenthos grew up nearby. The site was on the inner shelf; whether and where the depth of water was 1, 2, or 20 m seems immaterial. There is no doubt that open sea existed toward the east where later earth movements have hidden or destroyed the evidence of contemporaneous sedimentation in probably deeper water.

The analysis of the fauna shows its marine character and suggests normal salinity and oxygenation and a temperate or warm regime. Mechanisms of fossilization were described by Wade (1968). The distinction between resistant and non-resistant bodies or parts of them is important for the interpretation of their remains. Most of the medusoids were non-resistant which means that they decayed before the covering sediments were compacted. The resulting fossils are mostly casts of their more or less decayed or compressed surfaces, composite moulds of several aspects of the body, or multiple casts made by several separate sand laminae. Counterparts are rare. Resistant animals are those 'which were able to support the overlying sediments until diagenesis had set them in relatively permanent moulds' (Wade 1968, p. 244). In the described circumstances of deposition, these moulds are external. Small sand wedges may accumulate under projecting tissues on the surface of some specimens. One can observe deformation before burial where overfolding, contraction, collision with other individuals, or shrinkage has occurred. There was some deformation during burial, but no tectonic deformation. Conditions of deposition led only very rarely to infilling of internal spaces. In many fossils only one side of the flat or flattened body is known. This, together with the often fairly coarse grain size of the sand makes preservation at Ediacara somewhat inferior to that of Late Precambrian Metazoa from other localities.

## 2.3 Composition of the fauna

The preservation of the fauna in the Ediacara Member is not as faithful in morphological detail as that at the famous occurrences of soft-bodied fossils in the Middle Cambrian Burgess Shale, the Upper Jurassic limestones of Solnhofen, or the Eocene Baltic Amber. The variety of components of the oldest known assemblage of Metazoa was obvious to its first observers but its classification in terms of the system of living invertebrates presented extraordinary difficulties. The lack of detail because of preservation in sediment which is relatively coarse compared with the extremely fine-grained rock matrix at the famous younger fossil localities is only one of them. Others are the distortion of the soft bodies of animals during fossilization, their partial decay, and various other accidents of preservation. These problems have to be taken into consideration when the fossil organism, within the original range of variability of the population, is the subject of classification, rather than the variety of existing accidental shapes. For this reason, unique specimens not comparable with any others were, as a rule, put aside, because their accidental features could not be distinguished from biologically significant morphological characters. Even when many seemingly specifically identical specimens were available, statistical treatment of biometrical data was unprofitable for diagnostic purposes because of uncontrollable distortion of the bodies during fossilization. The process of casting and moulding of bodies on bedding planes meant that in many instances a truly three-dimensional view of the animals could not (or not easily) be obtained. Some of their soft bodies are so strongly distorted or decayed that even measurements of their length and width tend to be meaningless. Identification of newly found fossils depends on comparison with known ones from the younger Phanerozoic, but few of these have their soft parts preserved. It then becomes necessary to compare them with preservable parts of living organisms. However, the older the fauna, the greater is the possibility that in the course of evolution the kind of organism represented by the fossil has changed beyond immediate recognition or has become extinct. In this case its classification depends on the existence of some living organism which is similar in a meaningful rather than in a formal sense, so that its relationship can be suggested on the basis of structural homology, rather than configurational or geometrical similarity or functional analogy. Emphasis was therefore placed on similarities with existing taxa. This was considered preferable to taking an agnostic position and erecting new high-level taxa of unknown affinities. It was not unexpected that the oldest known assemblage of Metazoa should prove to be, in a sense, entirely new. Its possible relations to later faunas, including living ones, were of greater interest than insistence on novelty. Concerning the results of these procedures, questions have been raised and doubts have been expressed about the placing of some species but no major changes in the assessment of the composition of the fauna have been made in the last 25 years.

The composition of the fauna as it is known at this time is shown in Table 1. It is listed and will be discussed under the headings Coelenterata and Coelomata, indicating main grades of the organization of the Metazoa.

*Coelenterata.* (*a*) Hydrozoa Chondrophorina. The chondrophores are represented in the living marine fauna by only two genera (*Porpita* and *Velella*). In earlier zoological classifications they were subordinated as an order Disconanthae to the Siphonophorida which form complex, floating colonies. Modern studies of their morphology and development have shown that these two genera are unrelated to the siphonophores but closely related to, probably derived from and therefore taxonomically subordinated

Table 1. *List of fossils from the vicinity of Ediacara (named species only) (see Figs. 2.4, 2.5, 2.21, 3.2)*

Coelenterata
  Phylum Cnidaria
    Class Hydrozoa
      Order Hydroida, Suborder Chondrophorina
        Family Chondroplidae
          *Chondroplon bilobatum* Wade
          *Ovatoscutum concentricum* Glaessner & Wade
        Family Porpitidae
          *Eoporpita medusa* Wade
    Class Scyphozoa
      Subclass Vendimedusae Wade (MS)
        Order Brachinida Wade (MS)
          Family Brachinidae
            *Brachina delicata* Wade
            *Ediacaria flindersi* Sprigg
          Family uncertain
            *Rugoconites enigmaticus* Glaessner & Wade
            *Rugoconites tenuirugosus* Wade
      Subclass uncertain
        Family Kimberellidae Wade
          *Kimberella quadrata* (Glaessner & Wade)
    Class Conulata
      Order Conulariida
        Family Conchopeltidae
          *Conomedusites lobatus* Glaessner & Wade
    Medusoid fossils of uncertain affinities
        *Cyclomedusa davidi* Sprigg
        *Cyclomedusa plana* Glaessner & Wade
        *Cyclomedusa radiata* Sprigg
        ?*Beltanella gilesi* Sprigg
        *Mawsonites spriggi* Glaessner & Wade
        *Medusinites asteroides* (Sprigg)
    Colonial Cnidaria, possibly related to Pennatulacea
      Family Pteridiniidae
        *Pteridinium* cf. *nenoxa* Keller
        *Phyllozoon hanseni* Jenkins & Gehling
      Family Charniidae
        *Charniodiscus arboreus* (Glaessner)
        *Charniodiscus longus* (Glaessner & Wade)
        *Charniodiscus oppositus* Jenkins & Gehling
        *Glaessnerina grandis* (Glaessner & Wade)
Coelomata
  Phylum Annelida
    Class Polychaeta
      Family Dickinsoniidae
        *Dickinsonia costata* Sprigg
        *Dickinsonia brachina* Wade
        *Dickinsonia elongata* Glaessner & Wade
        *Dickinsonia lissa* Wade
        *Dickinsonia tenuis* Glaessner & Wade

Table 1 (*cont.*)

  Family Sprigginidae
   *Spriggina floundersi* Glaessner
   *Marywadea ovata* (Glaessner)
 Phylum Arthropoda
  Superclass Trilobitomorpha (or Chelicerata)
   Family Vendomiidae
    *Praecambridium sigillum* Glaessner & Wade
  Superclass uncertain
   Family Parvancorinidae
    *Parvancorina minchami* Glaessner
 (Systematic position unknown)
   *Tribrachidium heraldicum* Glaessner

or at least equated in rank to one of the suborders of the Hydrozoa. *Porpita* is radially symmetrical while *Velella* has an axis of symmetry through the centre of the oval body, and a sail which extends diagonally across its surface. Under the surface is a chambered, chitinous float. This is preserved in fossils from Ediacara. *Eoporpita* is a radially symmetrical medusoid form, with various kinds of polyps below, which may be preserved by infilling with sand (Wade 1972*a*). Two other genera, *Chondroplon* (Wade 1971) and *Ovatoscutum*, are known as casts or external moulds only. In the former the chambers of the float are partially infilled with sand. Bilateral symmetry which characterizes them is not unexpected in a swimming organism, even if driven by a sail-like dorsal extension of its dorsal surface or by a 'mantle' flap. The significance of the discovery of Chondrophora in the Ediacara fauna is, firstly, the proof of the existence of Hydrozoa in the fauna, secondly the ecologically important fact that Chondrophora float at the surface of the sea and exploit its phototrophic phytoplankton. The small size and fragility of most other hydrozoan polyps and colonies and the difficulty of preservation and recognition of hydromedusae would have impeded their recognition in the fauna of the Pound Quartzite. Their representation by highly specialized surface floaters indicates not only a lengthy evolution from unknown, probably small, benthic, less specialized ancestors, but also the habitability of the surface waters, with exploitable plankton and normal oxygen availability and normal salinity. The flux of ultraviolet radiation could not have been at harmful levels.

(*b*) *Scyphozoa.* The new subclass Vendimedusae was established by Wade (in press) for the best known medusae from Ediacara. They are characterized by a mouth which either lacked a typical manubrium, or had a small, conical extension or lobes separated by branching slits. The round gastric cavity extended into numerous branching radial canals. The gonad is claimed to be a lobate annulus in the outer disc, communicating with the canals. Numerous fine tentacles, where present, arose on the subumbrellar edge. There is no trace of a tetrameral symmetry. In the best known species, *Brachina delicata*, for which a new family and order are being proposed by Dr Wade, there is a single coronal furrow around the central disc. *Ediacaria*, which reaches a diameter of some 50 cm and has a similar furrow, may belong to this family. The genera *Hallidaya* and *Rugoconites* (with reticulating radial canals) may also belong to the Vendimedusae which Dr Wade considers as (not necessarily monophyletically) derived from an unknown stem group of the Scyphozoa. The Brachinida left no known descendants.

The recognition of the Vendimedusae as a distinct subclass of primitive Scyphozoa leaves only the unique genus and species *Kimberella quadrata* as a possible early representative of the living order Cubomedusida in the subclass Scyphomedusae. This probably tetrameral medusa is rare and still incompletely known. It shows complex internal structures interpreted by Dr Wade as indicating possibly cubomedusoid affinities. Its almost prismatic shape when fossilized has produced only lateral views, making interpretation difficult. The Cubomedusida (as class Cubozoa) are considered by Werner (1975, 1976) as being closer to the Hydrozoa than the other Scyphozoa (see Chapter 3).

The conularids (subclass Conulata) are considered as polypoid forms, with a chitinized epidermis and tetrameral septation. They show homologies with the juvenile Stephanoscyphus stage of the living Coronatida and resemblance to the scyphopolyps of other Scyphomedusae (Werner 1966, Glaessner 1971*b*, Bischoff 1978*b*). There can be no doubt about the close relationship of the Ediacarian *Conomedusites* with the Ordovician *Conchopeltis*. I find the derivation of the elongate Conulata which survived to Triassic time from the low-conical Conchopeltida convincing, though others have questioned this evolution.

A large number of other medusoid fossils, in fact the most common specimens found at Ediacara have been assigned to three or four named genera or, in the absence of distinctive characters, they were left unassigned. Despite these taxonomic difficulties, these medusoids of uncertain affinities deserve careful consideration and continuing study in close contact with the developing knowledge of other fossil and living medusae. The commonest medusoids in the whole fauna are those collectively referred to as "*Cyclomedusa*" (Wade 1972*a*, p. 205). These medusae are characterized by a circular outline and a variable number and spacing of concentric grooves. Most of them show also finer or coarser radial furrows. There is clear evidence of flexibility, either in obvious effects of vertical compression of a formerly conical centre, or in mutual lateral distortion of adjacent specimens. In all specimens the exumbrellar side is preserved as external moulds in downward position (convex hyporelief). Subumbrellar structures are not clearly recognizable. From the known facts, Wade (1972a, pp. 205-7) has developed extensive speculations about the morphology, phylogeny and mode of life of *Cyclomedusa*. This genus and its named species have been cited from many localities in other parts of the world but in fact most species are difficult to distinguish even in material from Ediacara. The great variability of the expression of radial structures suggests some influence of their state of preservation. These structures, which are not present in all species, must be distinguished from irregular radial splits caused apparently by desiccation and possibly also by compression. The concentric rugae, which are also variable, did not necessarily diminish the flexibility of the umbrella and its inferred pulsating movements. The great abundance of identical cyclomedusae on some bedding planes (see Wade 1972a, Pl. 41, Fig. 1, and a reference to 49 specimens on a bedding plane, p. 206) suggests their ability to swim as modern medusae do, and their occurrence in dense swarms, often drifting toward the shore in tidal currents. The greater probability of settling under water with the exumbrellar side downward was demonstrated by Wade (1968). There is no convincing evidence of attachment in this position but some living medusae feed on the bottom, resting on their exumbrellar surface. Until subumbrellar structures are clearly recognized, speculations on the systematic position of *Cyclomedusa* must remain inconclusive. What is known about this genus is compatible with Scyphozoa; there is no clear indication of affinity with hydromedusae.

A similar conclusion can be reached concerning *Medusinites asteroides* and *Mawsonites spriggi*. However, these species have distinct coronal furrows and can therefore be placed, at least tentatively, in the vicinity of *Ediacaria*, possibly representing Vendimedusae. No further specimens of the form described as *Lorenzinites rarus* have been found. I am inclined to consider it as an aberrant *Rugoconites*, despite original reservations (Glaessner & Wade 1966, p. 609) and place it in the synonymy of *Rugoconites enigmaticus*. It illustrates the undesirability of naming a solitary specimen; it has in fact already proved misleading. *Pseudorhizostomites howchini* is a common fossil which was considered to represent the marks of decaying matter escaping through the overlying sand lamina. Confluent radial grooves are preserved on its lower surface. With the discovery of *Rugoconites tenuirugosus* it has become very probable that this was the originator (compare Glaessner & Wade 1966, Pl. 103, Fig. 2, with

Wade 1972a, Pl. 43, Fig. 6). Fedonkin (1981a, p. 16) compared *Pseudorhizostomites* with the hydrorhizae of the living *Campanularia* but it is more likely that their upright stems would have been flattened by the approaching sand wave rather than being left standing in it.

Many medusoids remain unidentifiable in the present collection and further taxa may be found among them. Specimens from the Vendian of northern Russia were assigned to *Protodipleurosoma* Sprigg by Fedonkin (1981a). Similar configurations occur at Ediacara but they do not resemble closely Sprigg's unique holotype specimen.

(c) Sessile colonial Cnidaria. Sprigg (1949, p. 73) mentioned that some of the fossils found at Ediacara were 'referable to algae' but did not describe them. Similar Precambrian frond-like fossils had been previously described from southwest Africa by Gürich in the 1930s and subsequently redescribed by Richter (1955) as coelenterates, by Ford (1958) from the English Midlands as possible algae, and again in great detail by Pflug in the 1970s as a phylum Petalonamae. Jenkins & Gehling (1978) have added much information and Jenkins, Plummer & Moriarty (1981, p. 75) concluded that the concept of the 'Petalonamae may involve the artificial amalgamation of quite distinctive classes or even phyla'. This review of components of the Ediacara fauna is not the place to discuss questions involving fossils from many localities which will be considered later in this chapter. I listed these frond- or leaf-like fossils (Glaessner 1979b, p. A96) under the heading Problematical Coelenterata, 'Petalonamae Pflug, 1970', a name accepted provisionally for a convenient grouping of interconnected taxa 'without defining its status in classification and nomenclature'. Under the same name and with the same intentions the Ediacara fossils of this kind were listed by Glaessner & Walter (1975, 1981). This was apparently misunderstood by later authors and therefore this name is not used here.

The four genera of fossils from Ediacara representing sessile colonial Cnidaria are provisionally grouped in two families, Pteridiniidae Richter, 1955 and Charniidae Glaessner, 1979. Members of the former are rare at Ediacara and not well known as to their growth and functioning. *Charniodiscus* Ford was redefined by Jenkins & Gehling (1978). It has a foliate structure with a median rhachis, a stalk and an attachment disc. Affixed to the leaf are lateral structures (polyp leaves) divided into ridges (polyp anthosteles). Straight, narrow grooves are interpreted as impressions of spicules which strengthened the stalk, rhachis and leaves. The entire structure shows close resemblances in details to representatives of the living order Pennatulacea (class Anthozoa, subclass Octocorallia) to which they were assigned (Glaessner & Daily 1959, p. 394). They were originally wrongly placed in the African genus *Rangea* Gürich of which at that time only the holotype specimen was known. Pflug (1970*b*) and Germs (1972*a*, 1973*a*) found and described additional specimens which proved the incompleteness and the misleading features of the fossilization of the holotype and which led to the establishment of a family Rangeidae Glaessner, 1979 (Pflug 1970*b*, p. 203, *nomen nudum*). It has been found so far only in the Nama Group. *Charniodiscus* is placed in the family Charniidae, which is widely represented in assemblages of Ediacarian fossils from other regions. The genus *Glaessnerina* Germs is similar to and possibly a junior synonym of *Charnia* Ford. It is rare at Ediacara.

The other genera which are also rare at Ediacara represent the family Pteridiniidae. The morphology and biology (growth and functioning) of these organisms are not yet clear. The elastic deformation of most specimens during fossilization and embedding is extreme and pervasive and has led the investigators to different interpretations. I have described single, leaf-shaped fronds with transverse, curved primary furrows separating convex ribs (Glaessner 1959, Glaessner & Daily 1959, Glaessner 1963). A group of fragmentary leaves from a massive quartzite below the main fossiliferous beds was described by Glaessner & Wade (1966). Secondary grooving, as between the anthosteles of the Charniidae, is rare and minute and was not seen in specimens from Ediacara, hence the position of the (hypothetical) polyps is uncertain. The leaves or petaloids form groups of five according to Pflug or three according to Jenkins & Gehling (1978). In *Pteridinium* there is a complex structure on each side of the midline but in *Phyllozoon*, which is described as a two-dimensional array, the median line is zig-zag shaped. No stalks or attachment discs are known. While there are characters common to Charniidae and Pteridiniidae, the assignment of the former to the pennatulaceans is better founded than that of the latter, at the present state of their knowledge. There are no grounds for excluding them from the sessile colonial Cnidaria.

*Coelomata.* (*a*) Annelida Polychaeta. Worm-like animals constitute the second largest group of fossils from Ediacara, about 25% of the specimens collected. Most of them clearly represent species *Dickinsonia costata*, which is broadly oval in outline, with up to 120 segments, and flat. Four other species are known, with 200–300 segments and ribbon-like bodies. They constitute a monotypic family Dickinsoniidae. They are unlike most living annelid worms but not without a parallel among them, the genus *Spinther* (Fig. 2.4 J) which lives as a parasite on sponges. The word 'parallel' is used here deliberately: *Spinther* is probably not a direct descendant of *Dickinsonia* but more likely the result of a course of evolution which was parallel in

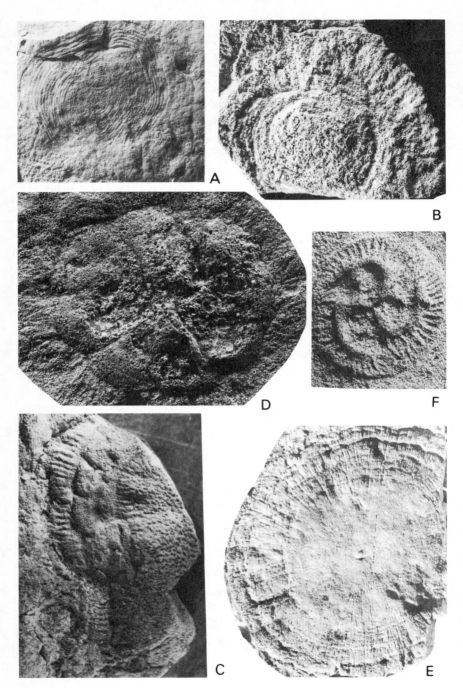

Fig. 2.4. Representative metazoan fossils from the Ediacara–Flinders Ranges area. A – *Ovatoscutum concentricum*, ×0.5. B – *Eoporpita medusa*, ×1. C – *Brachina delicata* (fragment), ×1.5. D – *Conomedusites lobatus*, ×1. E – *Cyclomedusa radiata*, ×0.5. F – *Tribrachidium heraldicum*, ×1.5.

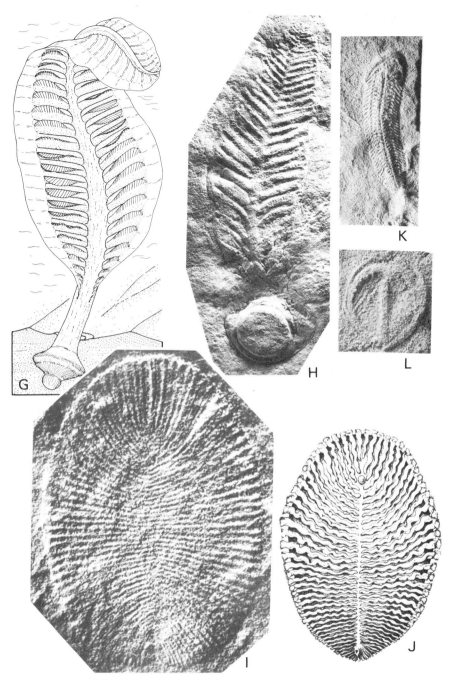

G – *Charniodiscus oppositus*, reconstruction, about ×0.4. H – *C. arboreus*, cast, ×0.23. (G and H from Jenkins & Gehling 1978). I – *Dickinsonia costata*, ×1. J – *Spinther alaskensis* (Recent, from O. Hartmann, 1948, *Pac. Sci.*, vol. 11). K – *Spriggina floundersi* ×1. L – *Parvancorina minchami*, ×1.

some respects but divergent in others. *Dickinsonia* provides a good example of what can be misunderstood and what can be reliably established in studying Ediacarian soft-bodied fossils. While only a handful of specimens were available and while the dominance of medusae in the fauna diverted attention from other kinds of animals, *Dickinsonia* was described as a possible coelenterate. Subsequently, a new class Dipleurozoa Harrington & Moore was erected for the family Dickinsoniidae. Important morphological characters were observed later. (1) At one end of the body the lateral extensions of the most anterior segments joined along the midline, while at the opposite end there was a small, triangular pygidium. (2) In two specimens infilled intestinal caeca could be seen which curved around what could have been a pharynx and then branched outward, while at mid-length they branched off laterally. (3) The intestine was straight and sediment-filled in some specimens. (4) The dorsal surface was distinguishable from the ventral representation of other specimens by a median interruption of the segmental boundaries. Notopodial ridges could be seen on either side of the median groove, folded either forward or backward and ending laterally in fan-shaped tips. (5) Less clearly, the possible presence of a prostomial tentacle on the dorsal surface was observed and the position of the mouth 'a short distance anterior to the U-shaped closure of the anterior body segment' deduced (Wade 1972*b*, p. 175). It is probable that there were no neuropodia. (6) The maximum known length of *Dickinsonia* was 45 cm. (7) Contraction by radial shrinkage, wrinkling, and folding of some bodies is known and contraction of longitudinal and transverse muscles is indicated. There is no evidence of damage from predation. A growth series of *D. costata* consisting of specimens from less than 4 mm to 170 mm length confirms that the species lived in the vicinity of its places of burial. This detailed evidence shows that these flat worms were not flatworms (Platyhelminthes) but had the essential morphological characters of annelid worms: segmentation with segmental grooving (and possibly additional wrinkling) of a cuticle, a digestive tract with caeca, notopodia. They had probably longitudinal and transverse muscles, and grew by the addition of segments. They were sediment and detritus feeders and as such they must have been able to crawl and to swim, at least in order to find suitable feeding grounds (Wade 1972*b*). (An important palaeobiological study by Runnegar (1982*a*) appeared too late for discussion. While adding much new knowledge, it left me unconvinced on points of disagreement.) Notosetae were certainly not 'readily preservable' in medium-grained sand. A recent statement to the contrary, leading to the assertion that 'There is no definite evidence that *Dickinsonia* was a polychaete and its actual affinities remain uncertain' (Conway Morris 1979*a*, p. 269) is surprising. So is the confusion of the definite, notopodia-bearing segments of *Dickinsonia* with the often-figured peristaltic waves of the ventral surface of the triclad land planarian *Rhynchodesmus* which form temporary transverse zones (myopodia) adhering to the surface during locomotion. This led Fedonkin (1981a) to assign dickinsoniid worms to the Platyhelminthes. The enclosure of the prostomium by the anterior joining of the first segments links *Dickinsonia* firmly with *Spinther*. Neither of these slow crawlers (or swimmers by vertical peripheral undulation) needed an anterior mouth, just as the (acoelomate, unsegmented) Turbellaria are still doing very well without it. *Spinther* kept its basic annelid neuropodia and developed claws at their ends to attach to and feed on sponges (of which there is no evidence in the Ediacarian faunas) while *Dickinsonia* fed on the bottom. It either had anterior sense organs or lacked them which would not be surprising in a microphagous crawler in a world without predators. While the evolutionary lines leading to *Dickinsonia* and (later) to *Spinther* could be described as parallel, they are certainly not convergent from totally different unrelated points of origin outside the polychaete annelids. Neither of these two genera can be considered as close to ancestral polychaetes, an assumption which was made once and refuted more than once.

The other bodily preserved polychaetes of the Ediacara fauna are of a very different kind. *Spriggina flounderi* and *Marywadea ovata* constitute the family Sprigginidae. In some characters they are so unlike any living annelid that some authors arrived at the conclusion that there is no firm evidence connecting them with the polychaetes, so that their affinities must remain unresolved (Conway Morris 1979a). Other observers, notably Birket-Smith (1981a), were impressed with their apparent similarities with arthropods. In view of these controversies and because of the reported occurrences of possible representatives of *Spriggina* on other continents (Germs 1973b, Fedonkin 1981a), the 25 best preserved specimens of *S. flounderi* from Ediacara and other localities in the Flinders Ranges have been re-examined, in cooperation with Dr Mary Wade, with the following preliminary results.

The body consists of a head (prostomium), a trunk with up to approximately 42 segments, and a very small pygidium. The head is horseshoe-shaped, with a smooth surface, a smoothly curved anterior margin, and pointed posterolateral projections. (In *Marywadea* it is half-moon-shaped, with much shorter projections.) Its width is about equal to that of the anterior trunk segments (including parapodia) but its median length is greater than that of the body segments. The median portion of its posterior margin is in most specimens sharply impressed and ∩-shaped.

It is probable that the front of the ∩ corresponds to the position of the mouth of the animal seen in its ventral aspect. In these specimens the bulging structure inside the rim is probably the equivalent of the 'buccal mass' of some living polychaetes. The two longitudinal rows of muscle blocks appear to diverge to its sides while in dorsal aspect they continue forward. A lump extending behind the head in some specimens, often displaced to one side of the median groove, appears to represent a pharynx. While the margins of the head are usually sharply impressed, they are commonly asymmetric because of overlap with the first body segment. The head was rarely fractured and never completely flattened by compaction. The asymmetry of the head makes it difficult to determine the precise shape of the first trunk segment. The body segments vary slightly in length according to their contracted or extended shape when embedded. The body was laterally flexible to the extent of 70° (in the holotype) to 100° angular curvature of its axis but it is, at the most, only slightly sinuously flexed or doubly curved. There is invariably a narrow median groove, with resistant angular blocks, probably representing longitudinal muscles, on both sides of each segment. Laterally the segments extend with a knee-like angle into what is taken to be parapodia. They end in some parts of some specimens in impressions of straight acicular setae. They are small and not easily photographed against the background of sand grain impressions and structural lineaments in the rock matrix.

The following comparison with living polychaetes and other invertebrates follow from these observations. The head of *Spriggina* differs from the prostomium of all living polychaetes in being very much larger and sclerotized. It is not divided by transverse or longitudinal grooves. It is without known appendages such as palps or antennae and no eyes are visible. *Spriggina* grows in width from about 3 to 10–11 mm until it has about 30–40 trunk segments. Only the length increases with further growth. The pygidium is small. The parapodia are similar to the neuropodia in living Aphroditidae. They are similarly close to each other and they could have been used in crawling with their terminal aciculae, in conjunction with lateral bending of the body. This would have required less elaboration of parapodial and acicular and more use of longitudinal muscles in locomotion. *Spriggina* was not a burrowing organism but was adapted to feeding on detritus on and in the soft mud in which it was preserved. As it had apparently no enemies other than Cnidaria, active predators being absent, it did not need elaborate sense organs, nor did it have to burrow for protection. As a whole range of growth stages from 2.5 to 46 mm length is found at Ediacara, the species apparently lived where it was fossilized. It was not an ancestral polychaete which according to Clark (1964) and

## Composition of the fauna

Fauchald (1975) would have been a burrower. There is nothing to exclude *Spriggina* from being one of such an ancestor's possible descendants, aberrant from others in its acquisition of a large head. This is a step in the direction of cephalization by expansion of the acron and sclerotization

Fig. 2.5. Comparison of Ediacarian and living polychaete worms. A – *Spriggina floundersi*, external mould of ventral side. Lighting arranged to suggest convexity. B – Latex cast of the same specimen. C – *Laetmonice producta* Kinberg, ventral view. Specimen in British Museum (Natural History), photograph reproduced by permission. Note buccal mass, longitudinal muscles, neuropodia. The felt-like dorsal covering projecting around the body is characteristic of the family Aphroditidae. Each scale bar, 10 mm.

of its integument but without incorporation of trunk segments. Otherwise it is unlike any arthropod in body and limbs. Among living polychaetes, the ventral aspect of *Laetmonice* (Fig. 2.5) resembles it closely in segmentation and in the setting of its neuropodia. The dorsal surface of *Spriggina* was not noticeably scaly or spinose but some specimens of *Marywadea* show long, curved, dorsal setae.

(*b*) Arthropoda. Two genera in the Ediacara fauna are assigned to the arthropods, on indirect but convincing evidence. The relatively small number of specimens complicates the necessary distinction between morphological characters and accidents of fossilization. Their small size obscures appendages. Where they do occur, their jointing is totally unrecognizable among the sand grains of the matrix. One of these genera, *Praecambridium*, is considered (Glaessner & Wade 1971) as an arthropod mainly on the basis of its segmentation and on the fact that the number of segments increases with its size, hence with growth. Distortion without fractures indicates that the body was without a hardened or mineralized integument. Articulation between segments or recognizable differentiation of arthrodial membranes cannot be expected. The segmentation resembles that of certain trilobite larvae and/or primitive Chelicerata. In the absence of traces of appendages which would have been minute and not detectable under the dorsal integument or even in composite moulds, speculation about possible taxonomic assignment to one or the other of these branches of arthropods would be fruitless, particularly in view of continuing controversies about their phylogenetic relations. The existence in the Ediacara fauna of an animal, however imperfectly known, which resembles primitive trilobites as well as chelicerates, is significant.

The other fossil, *Parvancorina minchami*, is assigned to the arthropods (Glaessner 1979*b*, 1980) because its remains consist of a shield-like, flat, unsegmented and unmineralized carapace, with traces of metameric ventrolateral appendages. They indicate segmentation of the body which does not necessarily conflict with the lack of its expression in the dorsal shield. Articulation of the podomeres, which need not have been fully developed in soft-bodied arthropods, would in any case be unrecognizable in their size range. Some of the range of their mobility can be seen in the varying position of the locomotory appendages in different specimens. No antennae are known. *Parvancorina* differs in essential characters from *Praecambridium* but it is comparable in some aspects with other primitive arthropods, of Middle Cambrian and Devonian age, and with the larval stages of branchiopod crustaceans. These comparisons (Glaessner 1980) lead to the conclusion that *Parvancorina*, as well as *Praecambridium* and related Vendomiidae (p. 85) can be placed 'near the point of branching of the ancestral Trilobitomorpha into Crustacea and Chelicerata'.

(c) *Tribrachidium*. The last of the named members of the Ediacara fauna, *Tribrachidium heraldicum*, cannot be definitely assigned to any known phylum. It is circular, disc-shaped and deformable. From its dorsal surface arise three fixed, angular arms with a fringe of short, soft, tentacular, movable projections on one side which turns to the periphery of the disc, and with a rounded organ on the other side. The dorsal surface also carries numerous, fine, stiffly flexible and possibly tubular spines. The mouth may have been in the centre of the disc but it is not clearly discernible. The arms and tentacles can be described as a lophophore; the threefold symmetry is unique. The only resemblance which has been commented on frequently is with edrioasteroid echinoderms in which one pair of three proximal ambulacra splits distally into two equal ambulacra each, resulting in the usual fivefold symmetry. However, there is no trace of the presence of calcareous plates on the surface of *Tribrachidium* prior to fossilization. Paul (1977, p. 126) expressed the opinion that 'Possibly *Tribrachidium*...is a primitive pre-skeletal echinoderm which resembles a soft-bodied edrioasteroid with only three ambulacra'. He derives also the Lower Cambrian Camptostromatoidea and Helicoplacoidea from this hypothetical progenitor but does not discuss its possible links with 'one of the minor coelomate phyla such as the Phoronidea' from which he believes the echinoderms have probably arisen. Derstler (1981) now describes *Camptostroma* as an edrioasteroid and the pear-shaped, stalked helicoplacoids as having three ambulacra winding from the mouth over the surface in different directions.

## 2.4 Palaeobiology

Before asking and answering, as far as possible, questions about the mode of life and interrelations of the elements of the faunal assemblage at Ediacara, its picture must be completed on indirect evidence. Firstly, we have to assume on general grounds the presence of microplankton. This would have consisted of a photoautotrophic microflora and of planktotrophic protozoans and metazoan larvae. Under the conditions of sedimentation of the fossiliferous rocks which have been described, no microorganisms could have been preserved. Secondly, several trace fossils occurring in these rocks (Glaessner 1969a) have to be considered. Traces of animal activity are fairly common but their diversity is significantly smaller than in Lower Cambrian strata at many comparable localities. Six different forms only have been recognized. None of them can be ascribed to animals known as body fossils. Each of them appears to have been made by a different animal: for reasons of size and the kind of activity represented, no grouping of them as different types of behaviour of the same kind of animal seems possible. Two obvious categories of trace fossils are hyporeliefs on the lower surfaces of sandstone slabs and epireliefs on their upper surfaces. The former corresponds to the common mode of preservation of body fossils, i.e. they represent casts and moulds of biogenic configurations on the surfaces of underlying layers of silt and clay. Two kinds of these configurations can be distinguished, three-dimensional objects such as chains of biogenic sediment pellets, and moulds of sinuous grooves. The chains of pellets (Glaessner 1969a, Fig. 5A), preserved as casts or moulds, can now be assigned to the form genus *Neonereites* Seilacher, 1960. (The names *Cylindrichnus* Bandel, 1967, *non* Toots 1967, and its replacement *Margaritichnus* Bandel, 1973 were subsequently found to be inapplicable; see Hakes 1976.) They can be interpreted as products of a sediment feeder which gathered sediment and passed it through its intestinal tract. The pellet formation may have been facultative, as segmented trails can pass into smooth ribbons. The sinuous grooves comprise two different kinds. One consists of small guided meanders covering areas of bedding surfaces (Glaessner 1969a, Figs. 5C,D). They are feeding trails of an organism capable of systematic, thigmotactic 'grazing' on microflora and organic detritus on the sediment surface. The other kind consists of gently sinuous, smooth, broad bands up to 25 mm wide (Glaessner 1969a, Fig. 5E). They are occasionally found to have cut through moulds of common Metazoa such as *Dickinsonia* which indicates that they were made after the basal lamina of sand had been deposited and hardened and therefore suggests an infaunal originator searching for food in the clay below the sand. Giving a name to a trace fossil which lacks distinctive morphological characters not related to local conditions would be of little use. Similar behaviour is suggested by a smaller, 5–6 mm wide trace fossil (Glaessner 1969a, Fig. 5F), the trail of which forms loops. Finally, there are three further different kinds. A common epirelief is an irregular, occasionally branching locomotion trail with raised 'levees' on both sides. These trails are 3–4 mm wide (Glaessner 1969a, Fig. 5B) and suggest an organism searching for food on the surface of a freshly deposited sand lamina or bed. The second kind consists of short, deep grooves with pointed ends, resembling the lower ends of the U-shaped burrows of *Corophioides*, as figured by Häntzschel (1975, Fig. 31, 5c), but there is no evidence of such burrows continuing upward. The third kind of surface impressions is rare. They consist of small, shallow holes, 2–3 mm distant, disposed at the corners of isosceles triangles, in negative hyporelief (impressed on lower bedding planes). This arrangement has not been found elsewhere and there is no interpretation for it at present. The conclusion from these palaeoichnological observations is that there were in the Ediacara assemblage, in addition to the bodily preserved fauna, at least six worm-like animals. They were sediment or detritus feeders which cannot be classified in the zoological system in which there are many different worm-like animals capable of leaving such traces.

A community of bottom-dwelling worm-like animals exploited the organic detritus and microbiota on and possibly in the sediment. The sediments were probably rich in organic matter but residual hydrocarbons were subsequently completely flushed out of the arenaceous, permeable sediments and expelled from the silts and clays. It is not known whether the producers of the described locomotion and feeding trails were acoelomates or coelomates but I would expect some of them to have reached coelomate grade, noting Clark's (1964) demonstration of the link between efficient locomotion and feeding and the acquisition of a coelom. The efficient grazing in thigmotactically controlled meanders presupposes relatively advanced sensory control of movement. Seilacher (1977) has shown that efficiency of that activity increased in later Phanerozoic time. There is at Ediacara no evidence of vertical burrows such as *Skolithos* which must have been made by animals with static organs informing them of vertical body position.

There is no evidence of intense bioturbation of deposited sediment, or of the existence of permanent dwelling burrows. The sediment/water interface was apparently exploited by rare sessile Cnidaria representing the Conulata (*Conomedusites*) and possibly the common *Cyclomedusa* which is incompletely known. The presumption that the medusoid Cnidaria fed on plankton and nekton with the aid of their nematocysts is hypothetic but highly probable. The annelids (Dickinsoniidae, Sprigginidae) were probably nektobenthic, swimming or crawling rather slowly in the detritus-rich bottom waters. The impression of low efficiency of locomotion and feeding conveyed by their known morphology, particularly when compared with living annelids, is probably correct. There was no need for rapid escape or protection from predators for which there is no evidence. Such protection is provided in some living polychaetes by burrowing or by curling up and 'presenting a battery of spines to any would-be aggressor' (Mettam 1971, p. 491). None of this was necessary in a community without macrophagous predators. The Vendimedusae which may have drifted toward the shallows of Ediacara in swarms from the open sea would have been armed with nematophores but none of them had the long mouth arms and tentacles of some living Scyphomedusae. In the absence of known pelagic competitors from other phyla they would have exploited the plankton biota. Another indicator of the presence of a rich source of food in the form of surface-seeking plankton is the chondrophoran Hydrozoa which are adapted by their chitinous pneumatophores to the

68    2: *The Ediacarian faunal assemblages*

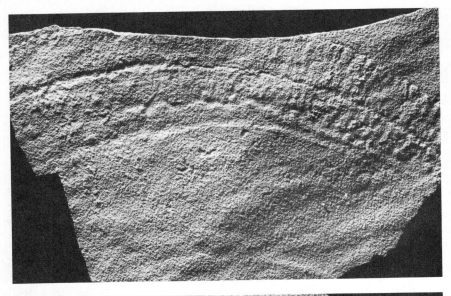

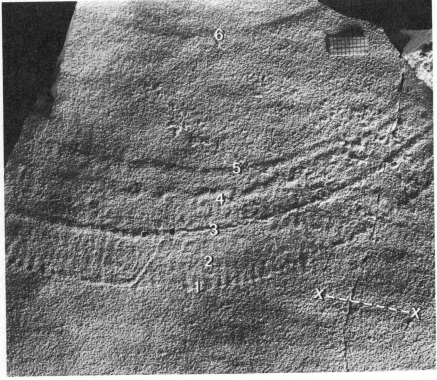

Fig. 2.6. For legend see opposite.

Palaeobiology

exploitation of the near-surface waters which are rich in phototrophic plankton. Finally, there are the sessile, colonial Cnidaria, probably related to the Pennatulacea (*Charniodiscus, Glaessnerina*), and the rare and not yet fully understood *Pteridinium* and *Phyllozoon*. Their polyps, like those of their living relatives, were small; they could not be preserved. They would have caught microplankton above the shallow bottom, at heights from above 10 cm to nearly 1 m. The astonishing size of some almost unprotected, soft-bodied, early Metazoa is obviously related to the fact that one of the main present marine trophic levels, that of macrophagous predators, was still unoccupied. Ediacara is the only known place where chunks of 'meat', 0.5 to 1 m in size, were once lying on the sea floor untouched until they decayed or were buried. This does not indicate scarcity or inefficiency of microbial consumers at that time or place. The occupation by Metazoa of habitable and food-providing niches in the sea was then still incomplete. We shall see (Chapter 4) that the slow but inexorable selection pressures eventually filled this gap, bringing the age of soft-bodied giants temporarily to a close. But first we shall review, in less detail, other Ediacarian fossil assemblages.

## 2.5 Other occurrences of Late Precambrian Metazoa in Australia

There is no need to discuss in detail the fossil content of other localities in the Flinders Ranges as it is exemplified sufficiently for the purpose of this study by the rich fauna from Ediacara. It is necessary, however, to mention fossils found some 2000 m stratigraphically below the Ediacara Member (Fig. 2.2) in the lower part of the Ediacarian Wilpena Group in Bunyeroo Gorge, in the dark silty shales of the Brachina Formation, or Moorillah Formation of the Brachina Group (Jenkins *et al.* 1981). First described (Glaessner 1969*a*) as a trace fossil *Bunyerichnus dalgarnoi* (Fig. 2.6), it was subsequently considered by Jenkins *et al.* (1981, p. 79) as a 'pseudofossil' but recognized again as the product of a living organism by Cloud & Glaessner (1982, p. 784). It is not a crawling track but most probably the subumbrellar imprint of a medusa, about 24 cm in diameter. It shows a distinctly scalloped, deeply imprinted outer edge with short ridges and grooves radiating inward, and several almost concentric

Fig. 2.6. *Bunyerichnus dalgarnoi*. The lower picture is of an upper bedding plane, the upper one is its counterpart on the bottom of the succeeding stratum. 1 – marginal lappets, 2 – radial canals, 3 – ring muscle or canal, 4 – zone of dominantly concentric and rare and obscure radial structures, 5 – inner edge of this zone, 6 – inner ring ridge (first noted by Dr R. F. J. Jenkins, personal communication). X – faint current lineations. Magnification: ×1.

ridges and grooves. The pattern resembles but is not identical with that of the radial canals and ring muscles of living rhizostomid medusae (Thiel 1978) and also the margin of the Ediacarian vendimedusid *Brachina* Wade, of which only exumbrellar and composite moulds are preserved. The deepest impressions being marginal, an outward pressure is indicated, as in an imprint of a flattened object, rather than an inward 'tethering' of an object swinging around in circles on a stalk. Another specimen found at the same locality is very closely similar to *Suzmites tenuis* Fedonkin, a mould of a problematic body, organic because regularly sculptured, from the Ediacarian of the White Sea coast (see p. 84).

Stratigraphically though not geographically related is another find (Fig. 2.7). The discovery by M. R. Walter of 'simple metazoan tracks in the Elkera Formation of north central Australia implies a...position at or beneath the base of the Bunyeroo Formation' by correlation between the Proterozoic in the Jervois and Flinders Ranges (Cloud & Glaessner 1982, p. 784 and footnote 21). This metazoan trace fossil is the oldest presently known in the Alice Springs region, being older than the fossiliferous Arumbera Sandstone and most of the Central Mt Stuart Formation.

Single fronds of *Charniodiscus*-like fossils have been recorded from other sedimentary basins in rocks of similar character and age. In the Officer Basin in the north of South Australia a single specimen of a large fragmentary frond (Fig. 2.8) was found in the Punkerri Sandstone (Major

Fig. 2.7. *Planolites* sp., Elkera Formation, Jervois Ranges, central Australia. Photo courtesy of Dr M. R. Walter.

1974), at 130°30′ E, 27°20′ S. Jenkins & Gehling (1978) have questioned the organic origin of this specimen which Daily had identified and which I consider as convincing. Other taxa named from this locality are not represented by recognizable specimens. Another unquestioned find of a similar but smaller specimen of *Charniodiscus* was made in the Amadeus Basin in central Australia, east of Deep Well, 80 km south-southeast of Alice Springs (Glaessner 1969a, p. 382, Fig. 9A). It was found loose in the talus of a low ridge consisting only of the lower part of the Arumbera Sandstone. Several separate and thorough searches at this locality failed to find other fossils. A medusoid fossil embedded in a partly decayed condition, was found higher (275 m above the base) in the Arumbera

Fig. 2.8. *Charniodiscus*? sp., Punkerri Sandstone, Officer Basin. Photo courtesy of Dr B. Daily. Magnification: ×1.

72    2: *The Ediacarian faunal assemblages*

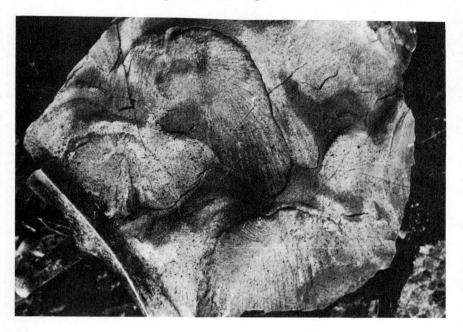

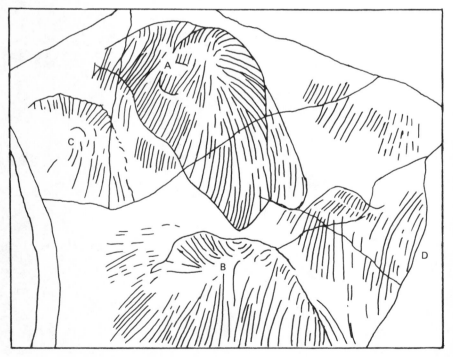

Fig. 2.9. For legend see opposite.

Sandstone 25 km southwest of Alice Springs and figured (Føyn & Glaessner 1979, p. 38, Fig. 8B) as *Kullingia* cf. *concentrica*. This horizon contains abundant, radially ribbed, apparently originally cup-shaped but mostly more or less flattened fossils (Fig. 2.9) described as *Arumberia banksi* Glaessner & Walter, 1975, from a number of localities near Alice Springs. Jenkins *et al.* (1981) questioned the organic nature of this fossil. The bedding planes are mostly modified by current lineations but the fossil is unlike any of the numerous examples of such lineations figured and described in the literature. The fact that similar configurations can be produced artificially (experimentally) by jets of water from a tap is irrelevant. Mechanical current lineations on bedding planes never radiate from hollows, as the ribs on many of the figured (convex hyporelief) specimens of *Arumberia* do, unless they are gas volcanoes with central vertical feeder channels which the Arumbera fossils do not possess. The same beds in the Arumbera Sandstone also contain the vendimedusid scyphozoan *Hallidaya brueri* Wade, 1969. This species and the medusoid *Skinnera brooksi* Wade, 1969 were also found in large numbers near Mt Skinner, about 180 km north-northwest of Alice Springs, in the Central Mt Stuart Beds of Late Precambrian age. This is at present the northernmost occurrence of Late Precambrian fossils in Australia. (A rock with *Skolithos* from the Wessel Islands off the north coast of Australia is now known to contain Cambrian trilobites, despite the presence of Precambrian glauconite in the same rocks. A supposed Early Proterozoic fossil from the Noltenius Formation of northern Australia was shown by Walter (1972) to be the result of sand volcano activity.)

In summary, it can be stated that Late Precambrian rocks of central Australia contain colonial cnidarians similar to those from Ediacara and also medusoid and other cnidarians which are not known from South Australia.

### 2.6 The Nama fauna of southwestern Africa (Namibia) and possible equivalents in South America

*Southwestern Africa.* The first discoveries of genuine Precambrian Metazoa were made in sediments from the basal part of what is now known as the Nama Group. Specimens were collected by P. Range and H. Schneiderhöhn as early as 1908 and 1914 and first described by G. Gürich in the 1930s. A long series of publications by Richter, Pflug, Germs and others has followed in the last 25 years and field and laboratory studies

Fig. 2.9. *Arumberia banksi*, Laura Creek, Amadeus Basin. Field photograph by the late J. Banks. (The sketch shows position of four specimens (A–D), drawn from another photo of this rock). (See Glaessner & Walter 1975, Fig. 1A). Magnification: ×0.33.

are still proceeding. Controversies about morphological and taxonomic interpretations are also continuing. Consideration of all named taxa at this time would not serve the purpose of this review (see Glaessner 1979b). Instead, the main constituents of this assemblage are indicated, mostly in terms of higher taxa, preceded by an account of their stratigraphic placing and followed by remarks on their preservation and palaeobiological significance.

The Late Precambrian fossiliferous sediments concerning us here are now classified in lithostratigraphy as the lower part of the Nama Group of the Damara Supergroup (Kent & Hugo 1978). The Nama Group forms the platform cover over an area (Fig. 2.10) roughly as large as the Flinders Ranges in South Australia. This platform is situated southeast of the Damara orogen (Martin 1965, Germs 1972a, Kröner et al. 1980). The Nama Group locally overlies the Numees Tillite with a slight unconformity.

The Nama consists of a sequence of alternating clastics and limestones of regionally varying thickness. It is divided into the Kuibis, Schwarzrand, and Fish River Subgroups (Fig. 2.11). These divisions, formerly considered by Germs as formations, are now subdivided into numerous named formations and members (Kröner et al. 1980, Tankard et al. 1982). Age limits of about 680–550 m.y. for the lower Nama are given by Kröner et al. The trace fossil *Phycodes pedum*, which is not known from the Precambrian, occurs in the Fish River Subgroup and in the uppermost part (Nomtsas Formation) of the underlying Schwarzrand Subgroup, without any other Nama fossil. It is concluded that the Fish River and Nomtsas units of the Nama Group are of Cambrian age while the rest of the Schwarzrand and the Kuibis Subgroups are Late Precambrian. The Nama assemblage contains a fossil which is confined to limestones in both the Schwarzrand Subgroup (Huns Limestone Member) and the Kuibis Subgroup (Schwarzkalk Member). This is a partly calcareous worm tube, *Cloudina hartmanae* Germs (1972a,b), of which I have described additional material (Fig. 2.12) confirming its assignment to the Cribricyathida and their relation to polychaete annelids which had been suspected by Germs (Glaessner 1976a). These tubular fossils were first described from the Cambrian of Siberia and erroneously considered as Archaeocyatha. The earlier appearance of a different species, genus and probably family of Cribricyathida in the Late Precambrian of southwestern Africa is not surprising but it has confused a number of observers. The Huns Limestone Member contains remains of medusoids, originally identified by Germs (1972a) as *Cyclomedusa davidi* Sprigg but actually closer to the medusoid genus *Tirasiana* Palij, 1976 from the Late Precambrian of the USSR.

Fig. 2.10. Distribution of Nama Group rocks in southwest Africa/Namibia. (From Tankard *et al.* 1982. For sections along north–south and east–west lines shown refer to this publication.)

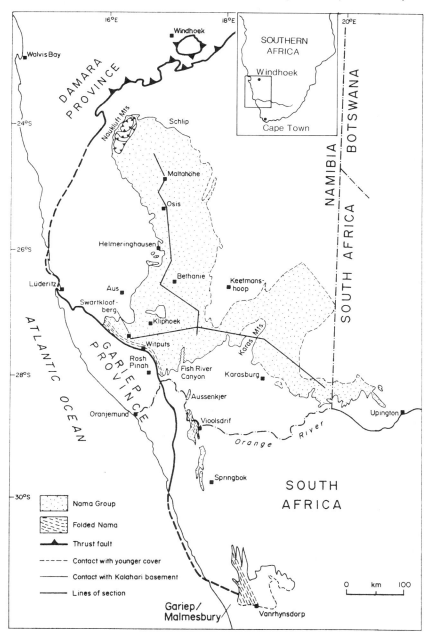

76   2: *The Ediacarian faunal assemblages*

Fig. 2.11. Stratigraphic column of Nama succession, as represented near crossing of section lines on Fig. 2.10. The column represents 2500 m of strata. Stratigraphic positions of fossil occurrences are indicated: B – body fossils, C – *Cloudina*, T – trace fossils. After Tankard *et al.* (1982) and personal communications from G. J. B. Germs.

| Fossils | Subgroup | Formation / Member | Description |
|---|---|---|---|
| T | FISH RIVER SUBGROUP | GROSS AUB | Shale and mudstone. Fossils: *Phycodes pedum* |
|  |  |  | Sandstone (very coarse grained, cross-bedding) |
| T |  |  | Mudstone with thin sandstone interbeds. Fossils: *Phycodes pedum, Skolithos, Enigmatichnus africani* |
|  |  | NABABIS – Haribes | Red arkosic sandstone (medium to coarse grained, channels, trough cross-bedding, asymmetric ripples); mudstone partings (desiccation cracks) |
| T |  | Zamnarib | Red mudstone with quartz arenite interbeds (channels, ripple cross-lamination, desiccation cracks, runzel marks) |
|  |  | BRECKHORN | Sandstone (trough cross-bedding, mudstone partings) |
|  |  | STOCKDALE | Red sandstone (conglomeratic and coarse grained basally, trough cross-beds directed to SE and E) and mudstone at top |
|  |  | VERGESIG – Niep | Red mudstone and sandstone overlain by red sandstone and mudstone in N; blue-green mudstone and limestone overlain by green sandstone and shale in S. Fossils: *Phycodes* and *Planolites* |
| C?,T |  | NOMTSAS – Kreyrivier | U-shaped valleys, striated floors, shale, diamictite |
| C | SCHWARZRAND SUBGROUP | Spitskop | Blue limestone (oolites, cross-bedded grainstones, columnar stromatolites, hardgrounds and desiccation cracks along E margin); interbedded sandstone, siltstone, shale. Fossils: *Cyclomedusa* |
|  |  | URUSIS – Feldschuhhorn | Green and red shale and siltstone with interbedded sandstone. Trace fossils |
| B,C |  | Huns | Blue and yellow limestone (oolites, first stromatolites) interlayered mudstones thicken eastward. Fossils: *Cloudina, Cyclomedusa* |
| B,T |  | Nasep | Arkosic sandstone (cross-beds directed to W overlain by mudstone-siltstone; interbedded limestone, quartz arenite and diamictite. Fossils: *Nasepia, Pteridinium, Paramedusium* |
| B |  | NUDAUS | Quartz arenite and arkose (polymodal cross-bed distribution), mud-pebble conglomerate, green shale (tidal bedding structures, desiccation features). Fossils: *Rangea, Pteridinium, Planolites, Skolithos* |
| C |  | ZARIS | Blue limestone (current-directed stromatolites), shale interbeds thicken to E. Fossils: *Cloudina* |
| B,T | KUIBIS SUBGROUP | DABIS – Kliphoek | Arkosic sandstone (planar cross-beds directed to W), flaser-bedded shale, quartz arenite, evaporite. Fossils: *Rangea, Pteridinium, Namalia, Ernietta, Orthogonium, Skolithos* |
| C |  | Mara/Kanies | Conglomerate, arkose (mudstone interbeds, cross-beds directed to W), shale, quartz arenite, organic-rich dolomitic limestone. Fossils: *Cloudina* |

Undescribed representatives of this genus occur in the Ediacara fauna. Underlying the Huns Limestone is the Nasep Quartzite. From this formation came the agglutinated worm tubes *Archaeichnium haughtoni* Glaessner, 1963, according to Germs (1972*a*, p. 208). They had been considered by Haughton as Archaeocyatha and assigned to the Kuibis Subgroup; while correcting his taxonomic assignment (see also Glaessner 1978) I had inadvertently followed his erroneous stratigraphic placing of this fossil. From the Nasep Quartzite came also (Germs 1972*a*) the medusoid *Paramedusium africanum* Gürich, the only specimen of which was unfortunately destroyed in the Second World War. The same formation also contains the leaf-shaped, thin-ribbed *Nasepia altae* Germs (1972*a*, 1973*a*) which is difficult to reconstruct as a functioning organism. The assemblage of fossils from the Schwarzrand Subgroup differs significantly from that of the underlying Kuibis Subgroup. An exception is *Cloudina* which is reported from limestone facies in both subgroups. *Pteridinium* is recorded from the Schwarzrand only as a questionable find, an identification which is not convincingly supported by the published illustration of this specimen (Germs 1972*a*, Pl. 21, Fig. 3).

The Kuibis Subgroup contains a much richer fauna but it has been found only at a limited number of localities, particularly in the vicinity of Kuibis. It includes *Pteridinium simplex* Gürich (Fig. 2.13), *Rangea schneiderhoehni* Gürich, *Ernietta plateauensis* Pflug and *Namalia villiersiensis* Germs. Each of these species represents a different family within a still undefined higher taxon (or more than one taxon) of sessile colonial Cnidaria. Pflug (1972) raised these families to classes within a phylum Petalonamae and named numerous lower taxa including 27 new species in the 'Erniettomorpha'. I have treated these families as 'Problematical Coelenterata' under the heading 'Petalonamae' Pflug, together with an uncertain family for *Nasepia* and *Namalia*, and expressed the opinion that many generic and specific distinctions are based on differences in growth and preservation (Glaessner 1979*b*). I reduced Pflug's thirteen erniettid genera to five, while Jenkins (in Jenkins *et al.* 1981, p. 71) considers that all Pflug's Erniettomorpha 'belong to a single genus and species, *Ernietta plateauensis*' (Figs. 2.14, 2.15 A–D). A further discussion of these taxonomic questions is beyond the scope of the present review. The wide range of remaining uncertainties despite much effort expended on collecting and describing fossils illustrates the need for more integrated biostratigraphy, sedimentology, taphonomy and palaeobiology in the study of Precambrian Metazoa from the Nama. A recent addition to the list of its fauna is the discovery of an early representative of the Echiura, *Protechiurus edmondsi* Glaessner, 1979*a* (Fig. 2.15E).

Fig. 2.12. *Cloudina hartmanae* Germs. Tube fragments exposed on a weathered surface of limestone. Lower Nama, Kuibis Subgroup, Zaris Formation; Quaggaspoort Farm, south of Bethanie, southwest Africa/Namibia. Sample from Anglo-American Company and Aquitaine–southwest Africa, by courtesy of Dr H. J. Oertli.
A,B – longitudinal sections showing dark, granulated outer layer and spindle-shaped sections through calcareous half-rings (right side of B); C – similar view of another tube; D – oblique section; E – two transverse sections; F – weathered tube surface with transverse ribs; G – tangential section through flattened tube fragment. Scale bars, 20 mm for F and G, 10 mm for A–E. (A,B,D,E,F partly also in Glaessner 1976*a*.)

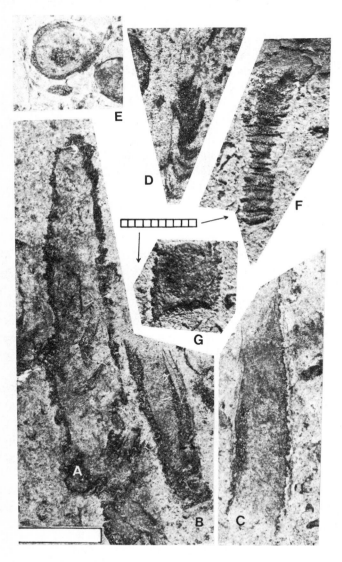

# Nama fauna of S Africa (and S America?)

The Kuibis fossils are found at some localities in numbers which are comparable with those of the Ediacara fossils but there are significant differences between the compositions and modes of preservation of these assemblages. Both are preserved mainly in quartzites. However, the bulk of the Kuibis fossils comes from massive quartzites in which the composite, sessile, leaf-shaped and bag-shaped organisms dominate, while medusoids and other floating or swimming animals appear to be absent. The massive quartzite was deposited as sand which enveloped the organic remains so that many of them are bodily preserved in three dimensions. At Ediacara their surface features were preserved as casts or moulds on bedding planes by mechanisms discussed above (pp. 49–50). The fossil which is closest to the dominant Kuibis organisms, *Pteridinium* cf. '*simplex*' was found 'in a

Fig. 2.13. *Pteridinium simplex* Gürich. Lower Nama, Kuibis Subgroup. A – × 0.5, B – × 1. Specimens in State Museum, Windhoek.

80    2: *The Ediacarian faunal assemblages*

massive quartzite' about 20 m below the main fossiliferous beds (Glaessner & Wade 1966, p. 617). This quartzite is believed to have been affected by later silcrete formation which is now known to have occurred regionally in mid-Tertiary time. Pflug (1970*a*, p. 233), however, states that the preservation of this material is due to synsedimentary silicification during the life of *Pteridinium* or soon thereafter. He mentions the preservation of organic matter observed by fluorescence microscopy but presents no

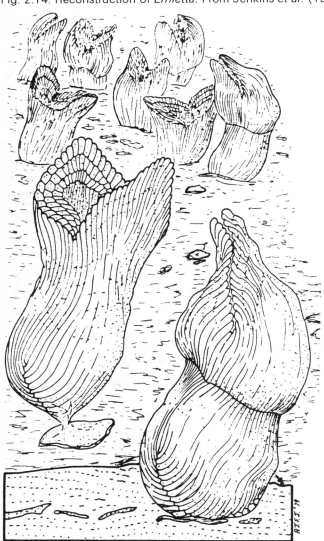

Fig. 2.14. Reconstruction of *Ernietta*. From Jenkins *et al.* (1981).

chemical analyses. Most of his material was collected on the surface as scattered blocks, weathered out of the rock. Effects of secondary silicification and weathering must be suspected. Such effects appear to have produced certain surface and internal (cavities and staining) features which enter into Pflug's elaborate reconstructions in the quoted and later publications.

The work of Pflug (1970a,b) and Germs (1972a, 1973a) has shown that while *Pteridinium*, one of the common members of the Kuibis association, occurs rarely at Ediacara, the genus *Rangea* is at present limited to the Kuibis Subgroup of the Nama. It is not adequately represented by the few specimens remaining from the early collections described by Gürich and Richter which were the basis for the erroneous placing of Australian species in this genus and subsequent references to it from other parts of the world. Rangeidae and Erniettidae *sensu stricto* are so far known only from the Kuibis Subgroup. A poorly preserved fossil from the same stratigraphic unit found about 30 km southwest of Kuibis was described by Germs as a 'sprigginid worm' but confirmation of this assignment by further specimens must be awaited. The occurrence of trace fossils in this unit proves that the fauna did not consist entirely of sessile organisms. Germs (1972a, c) refers to *Planolites* from the uppermost Kuibis Subgroup (the fossil figured as *Skolithos* does not appear to represent this Palaeozoic genus). The Nasep Quartzite of the Schwarzrand Subgroup is richer in trace fossils; about five different forms have been reported by Germs. Only the uppermost part of the Schwarzrand and the overlying Fish River Subgroup contain trace fossils known from the Cambrian.

*South America*. It has been suggested (Dalla Salda 1979) that the deposition of the Nama Group could have continued into South America, in the absence of an Atlantic Ocean in Late Precambrian time. Unfortunately, those Precambrian rocks in Brazil which would have been nearest to southwestern Africa are metamorphosed and unfossiliferous. However, Fairchild (1978) has now recognized a supposedly algal fossil named earlier (Beurlen & Sommer 1957) *Aulophycus lucianoi*, from the Tamengo Formation of the Corumbá Group in the State of Mato Grosso in Brazil, as possibly belonging to *Cloudina*. His investigations are continuing. In Argentina, in the Tandilia Hills which trend inland from the coast at Mar del Plata south of Buenos Aires, Dalla Salda and others correlate the La Tinta Group with the lower part of the Nama Group. No animal fossils have been described from the lower unit of the La Tinta Group, the Sierra Bayas Formation which is reported to be 700–780 m.y. old. The transgressively overlying unit, now known as the Balcarce Formation,

82     2: *The Ediacarian faunal assemblages*

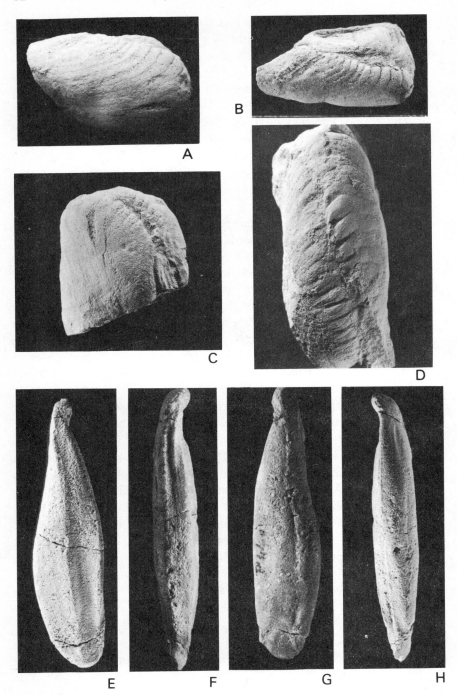

Fig. 2.15. For legend see opposite.

contains numerous trace fossils (Borello 1966). Borello separated tracks and tube fillings from 'fossil bodies' in which category he included *Bergaueria*-like casts and conical burrows, probably irregularly spiral, which are not known elsewhere. They are not body fossils in the usual sense. The age of the trilobite tracks and other trace fossils in these assemblages is Lower Cambrian to Ordovician. There are no distinctive Precambrian forms among them but this does not preclude the correlation of part of the La Tinta Group with some of the Nama Group or the possibility of future discoveries of Kuibis fossils in the La Tinta.

## 2.7 Late Precambrian Metazoa from the Northern Hemisphere

There is no firm consensus on the question of whether the Phanerozoic land masses of Eurasia and Laurentia existed in some form in Late Precambrian time as parts of a supercontinent or as two continents, or whether they were separated as a multitude of continental nuclei, occupying different positions and subsequently assembled by plate tectonic movements. The question of whether palaeobiogeographic constraints on such positions and movements follow from palaeontological data will be examined later (Chapter 5). This question should be kept in mind while their present geographical locations are reviewed. They are here presented in a sequence from the richest and best known to the poorer and less well known local assemblages, for reasons of ease of presentation and appreciation of the data.

(1) USSR. At the present stage of exploration, collecting, description and publication of Ediacarian faunas, the greatest number of specimens and taxa is known from the Late Precambrian sediments which form part of the cover of the East European Platform. Smaller numbers of finds are known from the Siberian Platform. Most of the pre-Palaeozoic platform cover of eastern Europe consists of terrigenous, sandy or silty sediments, without significant carbonate deposits. The latest Precambrian unit, the Vendian (Sokolov 1952, 1958, 1973) includes, or overlies according to another definition, widespread glaciogenic sediments and, in some areas, volcanics and tuffs (Volyn Series). The youngest Precambrian strata, long known as Gdov Series and Laminarites Series, are now named the Valdai Series. Richly fossiliferous localities in this unit (Keller & Fedonkin 1976)

Fig. 2.15. A–D – *Ernietta plateauensis* Pflug. Specimens in State Museum, Windhoek. A–C, No. P.1.16, figured by Pflug (1972, Pl. 34, Figs. 1–3, 6); B – view from upper right of A, C – view from lower right of A. D is another specimen, for comparison with *Pteridinium*. E–H – *Protechiurus edmondsi* Glaessner. Lower Nama, Kuibis Subgroup. All ×1.

were discovered in the 1970s in northern Russia, about 60 km west and later also about 100 km north of the town of Arkhangelsk (Figs. 2.16, 2.17). A few years earlier the first single finds were reported from bores in northern Russia and from outcrops on the Dnjestr River in Podolia, on the western flank of the Ukrainian Shield, some 300 km north of the Black Sea. Collecting, description and analysis of these faunas which are increasing steadily in numbers of specimens and taxa, are proceeding. A complete evaluation must therefore await the completion of these investigations. Although the distance between the localities in the north near the Arctic Circle and in the south near the Black Sea is about 2200 km, the assemblages are similar. This is less surprising than the fact that they have about 10 genera and 10 species in common with the antipodean locality of Ediacara. M. A. Fedonkin, who is describing the fauna (Fedonkin 1978*b*, 1980*a,b*, 1981*a*), believes that his present collections may contain 60–80 species of body fossils, in addition to trace fossils (personal communication).

The delicate, trilobed *Albumares* was placed in the Scyphomedusae (Fig. 2.18A). Other new medusoids are poorly preserved; some bear tentacles. *Albumares* resembles *Skinnera* Wade, and *Nimbia* resembles *Hallidaya* but lacks traces of 'central bodies'. Medusoids together with *Pteridinium* (Fig. 2.18B), *Charnia* (Fig. 2.21B) and *Charniodiscus*-like and other sessile organisms dominate the faunas of the White Sea coast. Of the species present, 69% are coelenterates. They are accompanied by *Dickin-*

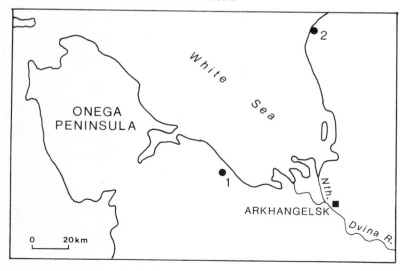

Fig. 2.16. Fossil localities in Upper Vendian (Ediacarian) of the White Sea coast, northern Russia.

*sonia* (Fig. 2.18D) and *Tribrachidium* (Fig. 2.18G). *Spriggina*-like fossils are rare but several primitive arthropods including *Vendia* (Fig. 2.18F) and *Vendomia* (Fig. 2.18E) as well as still undescribed multisegmented forms and possibly also *Parvancorina* occur. Some medusoids are assigned to the Hydrozoa, but without very convincing evidence. Others (*Nemiana* Palij and *Tirasiana* Palij) are considered as casts of polyp bases. This possibility has been considered for Ediacara fossils but is difficult to prove. *Pseudorhizostomites* is compared with hydrorhizae but this interpretation is unlikely, at least for the specimens from Ediacara. *Pteridinium* cf. *simplex*

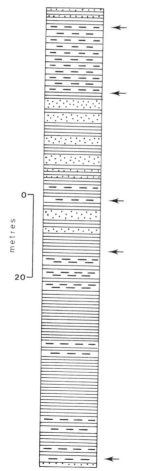

Fig. 2.17. Stratigraphic column showing position of fossiliferous strata at locality 2 (Fig. 2.16), 'Winter Coast'. (Lines – clay, broken lines – silt, dots – sand.) After Fedonkin (1981*b*).

86   2: *The Ediacarian faunal assemblages*

Fig. 2.18. Selected fossils from the Ediacarian fauna of the White Sea coast, northern Russia. A – *Albumares brunsae* Fedonkin, ×2.6. B – *Pteridinium nenoxa* Keller, ×1. C – *Inkrylovia lata* Fedonkin, ×0.8. D – *Dickinsonia costata* Sprigg, ×1.

from Ediacara is placed in the synonomy of *P. nenoxa* Keller. It also resembles the flat-surfaced *Inkrylovia lata* Fedonkin (Fig. 2.18C). The apparent difference between *Pteridinium* and *Inkrylovia* may be specific (regarding the surface) and preservational (regarding the margins) rather than generic. A remarkable specimen assigned to *Charnia masoni* Ford (Fig. 2.21B) shows clearly preserved anthosteles subdividing the secondary branches which Fedonkin describes as zooids. The figure (2.21B, compared with A, C, and D) suggests that in the diagnostic character, the angle between primary and secondary grooves, it is intermediate between the holotypes of *Charnia* and *Glaessnerina* which therefore may be a junior synonym of *Charnia*. This taxonomic matter requires further study.

The assignment of *Dickinsonia* and *Palaeoplatoda* to the Platyhelminthes is unacceptable because of the clearly polychaete organization of the Dickinsoniidae. These are relatively minor (though not unimportant) matters of interpretation and classification which in no way reduce the significance of Fedonkin's remarkable contribution to our knowledge of the Vendian fauna. The trace fossils and their significance will be discussed later (p. 183).

Much of this fauna is also reported from Podolia in the south (Palij, Posti & Fedonkin 1979, Fedonkin 1981a, p. 36, footnote). One of its medusoids, *Tirasiana*, was also described (Bekker 1977) from the sequence of strata of Vendian age on the western slope of the central Ural Mountains. *Arumberia* was found about 1000 m higher in the same sequence (personal communication from Yu. Bekker 1979). A tip of a frond which closely resembles specimens of *Charniodiscus* from Ediacara was recorded by Sokolov (1976a,b) as *Charnia* ex gr. *masoni* Ford from a bore at Nizhnaya Pesha, at a longitude of about 48° E near the coast of the Barents Sea, west of the Timan Range, just north of the Arctic Circle, at a depth of 3800 m. A unique find was recorded by Sokolov from another bore at Nepeitsino northeast of Moscow, from a depth of 1417–1428 m (Vendian, Redkino Series). The fossil (Fig. 2.19), named *Redkinia spinosa*, consists of several separate, dark brown (chitinous?) elements, about 3.5 mm long, comb-like, with curved and minutely spinose teeth on the convex side of a curved, flat base. Sokolov considers them as Protonycho-

Caption for Fig. 2.18 (*cont.*)
E – *Vendomia menneri* Keller, × 8.3. F – *Vendia sokolovi* Keller, × 4. From bore at Yarensk. G – *Tribrachidium* cf. *heraldicum* Glaessner, × 4. (Note that the surface is covered with thin, flexible spicules or tube feet which are only rarely preserved in specimens from Ediacara.) Photographs presented by M. A. Fedonkin and B. S. Sokolov.

phora. They could be arthropod remains but also resemble large scolecodonts (annelid jaws) while similarity to conodonts is very remote. Sokolov and many of his colleagues place all Vendian Metazoa from northern Russia and from Podolia in the Redkino Horizon of the Middle Vendian, while others correlate them with the Upper Vendian Kotlin Horizon. These correlations were reported by Fedonkin (1981a) who also indicates that the potassium–argon age of the glauconite in these strata was determined as 600 m.y.

In Siberia, the equivalents of the Vendian are mostly limestones and dolomites. In northern Siberia (Olenek Highlands, west of the mouth of the River Lena) *Glaessnerina sibirica* (Sokolov) was found in limestones. Many other Ediacarian Metazoa were collected there in 1981 (personal communication from M. A. Fedonkin and B. S. Sokolov). *Cyclomedusa plana* Glaessner & Wade from the River Maya near the Sea of Okhotsk, *Cyclomedusa* sp., *Pteridinium*, the *Namalia*-like *Baikalina* and trace fossils were figured and (very briefly) described from near Lake Baikal (Sokolov 1972b,c, 1973). The enigmatic large fossils *Suvorovella* and *Majella* were described from the River Maya region in eastern Siberia (Vologdin & Maslov 1960) and are mostly considered as algae but Sokolov (1972b, 1976b) recognized a resemblance of *Suvorovella* to some Ediacara fossils such as *Brachina* Wade and insists on placing them among the medusoids.

In view of the pronounced facies differences between the terrigenous Vendian of the East European Platform and its dominantly carbonate equivalents (Yudomian) on the Siberian Platform and on its margins, it is not surprising that differences in the faunas and their abundance exist. The correlation of the Late Precambrian in these regions is controversial, particularly that of the Vendian–Cambrian and the Yudomian–Cambrian boundaries (see Chapter 4). Biostratigraphic data play an increasingly important part in these discussions.

The Sabelliditida, narrow, thin-walled tubes of flexible organic material,

Fig. 2.19. *Redkinia spinosa* Sokolov, × 20. From a bore at Nepeitsino, Moscow Syneclise, Vendian. From Sokolov (1976b).

occur in considerable numbers and some diversity of form and structure in the lowest Cambrian of the Baltic area and in Scandinavia. Sokolov consistently considers them as tubes of Pogonophora, a peculiar group of worms of uncertain systematic position (see p. 121). Of six genera, two (*Paleolina* and *Calyptrina*) are recorded from both the Baltic basal Cambrian and the Late Precambrian of Siberia (Sokolov 1967, 1968, 1972*a*). Two occurrences (Fig. 2.20) are in the Vendian equivalents on the

Fig. 2.20. Tubes of Sabelliditidae. A – *Paleolina evenkiana* Sokolov, Upper Vendian, River Sukhaya Tunguska, northern Siberia. Macerated specimen in glycerine. B – possibly same species from Norilsk, northern Siberia. Preserved as carbonaceous film on black shale in bore core. Width of tubes about 1 mm. From Sokolov (1973).

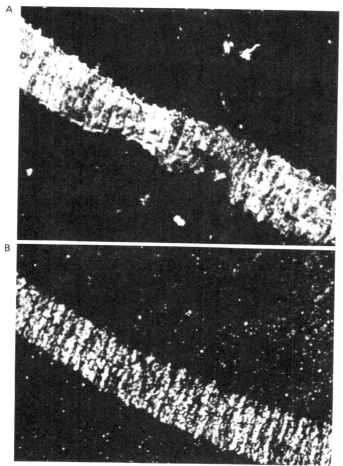

northwestern margin of the Siberian Platform (Lower Yenisey River) and another one is near Lake Baikal. Occurrences in the Upper to Middle Riphean of Siberia (about 950 m.y. old) have also been reported. They have been questioned, as have been reported occurrences of the same age in China.

(2) China. The first finds of metazoan fossils in the Late Precambrian (Sinian) of China were announced only recently. They are being studied at present and only preliminary information based on recent Chinese publications and on personal observations in the field and in collections can be given. An important, recently published collection of papers on the Sinian (Wang *et al.* 1980) contains important stratigraphic data and illustrations of fossils. References to *Cyclomedusa* from the top of the Changlingzi Formation (but formerly assigned to the base of the overlying Nanguanning Formation) in southern Liaoning have proved on re-examination in the field and in the laboratory (B. Daily and M. F. Glaessner 1979, unpublished) to be abiogenic configurations on upper bedding planes of silty limestones. These pseudofossils were caused by rhythmic gas eruptions which left channels, now calcite-filled, in the underlying matrix and concentric rings or occasionally radial markings on their upper surfaces. Liu Xiaoliang (1981) of the Shenyang Institute of Geology and Mineral Resources has reported *Arumberia* or *Baikalina*-like fossils and *Glaessnerina* from the Mashan Group, or Hsiatung Group of the Sinian Geochronologic Scale (Chung 1977, 700–570 m.y. old), of Longshan, northeast China. A *Charnia*-like fossil (Ding & Chen 1981) occurs in the middle Dengying Formation (Vendian equivalent) of Shipantan, Yangtze Gorge, Hubei Province. A specimen is in the collection of the Wuhan College of Geological Science. From the top of this formation *Anabarites* was reported. Tubes of *Sabellidites* also occur there.

(3) Scandinavia. Returning to the West, we note the find of three specimens of a medusoid fossil on the northwestern flank of the Baltic Shield in Sweden (Kulling 1972) which was first described as *Cyclomedusa*. It differs significantly from this genus and was named *Kullingia concentrica* Glaessner (in Føyn & Glaessner 1979). Detailed regional stratigraphic correlation places it in the uppermost Vendian, equivalent to the lower Vardal Sandstone of the Mjösa area in southern and the upper Stappogiede Formation of Finnmark in northern Norway. The trace fossils in these formations are described as simple burrows. *Rusophycus* and other trace fossils which are known from the Lower Cambrian occur together with *Platysolenites* in the next higher Breivik Formation (Banks 1973). The occurrence of a basal Cambrian zone with *Sabellidites* which was proposed from bore data south of the Baltic Sea and expected from finds of surface

Fig. 2.21. A – *Charnia masoni* Ford. Plaster cast of holotype from Charnian of Leicestershire, England. B – *C.* cf. *masoni* from Vendian, White Sea coast. C – *Glaessnerina grandis* (Glaessner), Ediacara. D – *Glaessnerina sibirica* (Sokolov), Olenek Highlands, Vendian, northern Siberia. Scale bar – 5 cm.

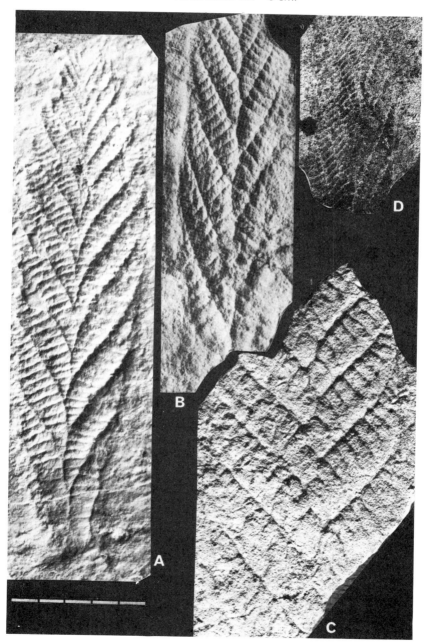

material in Finnmark, has now been proved in northern Finland, on the margin of the Caledonides (Tynni 1980).

(4) England. Important discoveries of an assemblage of Precambrian fossils were reported from the Charnian rocks of Charnwood Forest near Leicester in the English Midlands by Ford (1958, 1963, 1968, 1979). The assemblage includes fronds of *Charnia masoni* Ford (Fig. 2.21A) and *Charniodiscus concentricus* Ford, and also medusoids resembling *Cyclomedusa* and *Kullingia* but not necessarily identical with either of these genera. There are also other poorly preserved remains in the fossiliferous strata. They are assigned to the Woodhouse Beds of the Maplewell Group of the Charnian. The rocks are described as cleaved and indurated tuffaceous siltstones. The tectonic deformation of these rocks during later orogenic events has affected some of the fossils, so that some medusae have elliptical rather than circular outlines. These events have also affected the isotopic dating of the Charnian (Meneisy & Miller 1963, Cribb 1975, Dunning 1976), which is particularly important because according to its interpretation the fauna can be considered as either Early or Late Vendian.

Meneisy & Miller obtained whole-rock potassium–argon dates from 'porphyroids' (dacitic intrusives) considered to postdate the Woodhouse Beds. They considered the oldest date obtained as most reliable, because of argon loss in others. 'The oldest Charnian date...came from a "porphyroid" on Bardin Hill' (about 6 km northwest of the fossiliferous localities). The date was $684 \pm 29$ m.y. However, later rubidium–strontium isochron dates (Cribb 1975) on the 'Southern diorites' intruding the fossiliferous Woodhouse Beds and probably related to the volcanicity during and after the deposition of the fossiliferous sediments gave an age of $552 \pm 58$ m.y. Rubidium–strontium isochron dating could not be carried out for the 'porphyroids'. Hence there is no confirmation for the often quoted approximate age of 680 m.y. for the Charnian. It rests on the potassium–argon dating of a single sample from what Cribb calls a separate rock mass, the Bardon Hill 'porphyroid'. The date given by Cribb as $552 \pm 57$ m.y. and recalculated by Beckinsale *et al.* (1981) as $540 \pm 57$ m.y. is in agreement with the rubidium–strontium date of $533 \pm 13$ m.y. obtained by Patchett *et al.* (1980) from a pre-Late Tommotian granophyre in the Wrekin area, some 100 km west of Charnwood Forest. Beckinsale *et al.* (1981, p. 72), giving further datings for the Malvernian igneous rocks about 120 km southwest of the Charnian outcrops, conclude that the isotopic ages 'seem to define an older igneous episode at c. 700–640 m.y.... and a younger episode at c. 600–540 m.y. which includes late Precambrian and Cambrian metamorphic and igneous activity...in... Central England'. The Charnian fossiliferous sediments may have been formed at some time between these events, i.e. between 640 and 600 m.y. ago, since no geological or geographical link with the earlier of these episodes has been demonstrated. However, the composition of the Charnian sediments suggests that they belong to the Precambrian part of the younger of these episodes. Hence, their age may be between 620 and 580 m.y. This suggested age agrees with the dating of other occurrences of *Charnia* and *Charniodiscus*.

Ford (1980, p. 82) has now placed the Charnian fauna cautiously in a 'time bracket from before $552\pm58$ (Cribb 1975) perhaps as far back as $684\pm29$ m.y.'. He comments that this 'places them in the Vendian part of Precambrian time' when comparable fossils occur elsewhere. Cope (1977) found medusoid fossils south of Carmarthen in south Wales and considered them as an 'Ediacara-type fauna' while recognizing that 'the possibility that the fauna is of Cambrian age cannot be excluded'. One of the figured medusae resembles the Cambrian genus *Velumbrella* while the other one with equally spaced rings on its disc is unlike any of the Ediacaran medusoids; another fossil was 'tentatively linked' with the Lower Cambrian trace fossil *Astrapolithon*.

The fossiliferous Charnian and the Cambrian Hartshill Formation of Warwickshire have been hypothetically linked sequentially (Fig. 2.22) in a manner which is compatible with Dunning's (1976) tentative correlation of the Precambrian of Nuneaton where volcanics are overlain unconformably by the fossiliferous Hartshill Formation. Brasier, Hewitt & Brasier (1978), Brasier (1979) and Brasier & Hewitt (1981) record trace fossils (*Arenicolites, Gordia, Planolites,* ?*Psammichnites* and others) from the base of the Hartshill Formation which contains an Early Cambrian (Late Tommotian or Atdabanian) fauna about 250 m higher, in its upper part.

(5) Newfoundland. A rich fauna which has affinities with the fossils from Charnwood Forest was found in 1967 in southeastern Newfoundland (Anderson & Misra 1968, Misra 1969). Regrettably, over 12 years later it has not been properly examined, described and figured. It occurs (Figs. 2.23, 2.24) in what is now known as the Mistaken Point Formation at the top of the Conception Group (Williams & King 1979). It is described as 400 m thick, thin- to medium-bedded grey to pale red sandstone and purple shale, with minor tuff, and at its type section 'profusely fossiliferous with a variety of frond-like and disc-like impressions. The Late Precambrian forms occur on bedding plane surfaces that commonly display asymmetric wave ripples. In all cases the fossils occur beneath dark, thin, tuff horizons that are less than 1 cm thick. Fossil horizons are most common toward the top of the formation and five or more fossiliferous horizons are easily identified at the type locality'. The impressions can be properly identified

## 94  2: *The Ediacarian faunal assemblages*

Fig. 2.22. Possible Charnian – Lower Cambrian succession in Midlands area, England. From Brasier *et al.* (1978). (The fossiliferous Home Farm Member containing the *Obolella groomi* fauna (Brasier & Hewitt 1981) is now correlated with the uppermost Tommotian *Mobergella* zone of northeastern Europe. The lowest indicated position of the top of the Precambrian is therefore probably correct.)

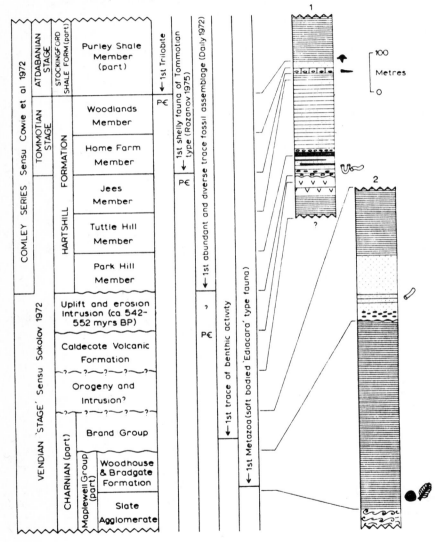

only from artificial moulds. An analysis of the branching of the 'spindle-like' impressions, the results of which differ significantly from the drawing F of Williams & King (1979, Fig. 14), was based on a mould kindly supplied by Dr Misra (Fig. 2.25). It suggested a new floating colonial hydrozoan of the order Thecata. The drooping bundles of similarly branching configurations described as 'dendrite-like' by Misra (1969, Pl. 6A, Pl. 8B) could well represent the same organism in a collapsed instead of spread-out presentation. It has not been found elsewhere but the resemblance of its small tubes to the terminal branches of *Rangea* from the Nama Group should be noted. There is no doubt about the occurrence of *Charnia masoni* Ford in this fauna, and a '*Cyclomedusa*' with numerous concentric annulations is said to resemble closely an impression from the Charnian described by Ford (1968). Disc-like impressions are probably of different origin. Some resemble basal discs of *Charniodiscus*, some appear to be floats of cnidarians, and some are basal organs of fine ribbed and stalked leaf-shaped organisms which apparently do not occur elsewhere (Misra 1969, Pl. 6A, Pl. 7A). The close biostratigraphic and lithostratigraphic

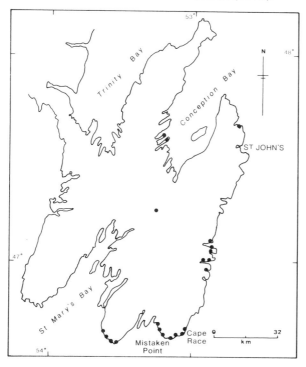

Fig. 2.23. Fossil localities in Upper Precambrian, southeastern Newfoundland. Localities from Anderson (1978).

96  2: *The Ediacarian faunal assemblages*

Fig. 2.24. Stratigraphic column of Upper Precambrian, southeastern Newfoundland. M – horizons with fossil Metazoa. Modified from Hofmann, Hill & King (1979).

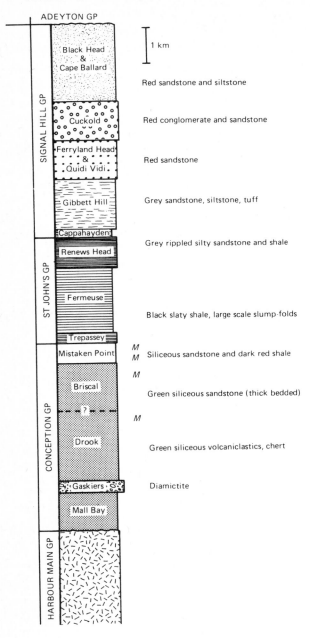

Fig. 2.25. Plastic mould of undescribed fossil (hydrozoan colony?) from Mistaken Point, Newfoundland, ×1. Photograph from S. B. Misra.

similarity of the Mistaken Point Formation and the fossiliferous Charnian has been pointed out by many observers and stratigraphic correlation is widely accepted. The occurrence of a tillite in the lower part of the Conception Group is significant for correlation with other fossiliferous Vendian strata but no tillites are known from the English Midlands.

The dating of the Conception Group has been the subject of much discussion (see Anderson 1972 and references therein). According to Anderson, the tillite of the lower Conception Group is about 670–715 m.y. old and the fossiliferous strata are dated 610–630 m.y. The Holyrood granite is thought to be younger than the Conception Group. Its rubidium–strontium isochron age is about 574 or $607\pm11$ m.y., depending on whether a decay constant of $1.39\times10^{-11}$ or $1.47\times10^{-11}$ is used. The recently adopted standard is $1.42\times10^{-11}$, resulting in an intermediate number (594 m.y., Cloud & Glaessner 1982). All these datings appear to be inconsistent with the concept of the Holyrood granite intrusion predating the Conception Group, as suggested by the table of formations in the paper by Williams & King (1979). However, there is agreement that the Conception Group can be correlated with the Vendian, the tillite representing the Varangian and the Mistaken Point Formation part of the Ediacarian (see Table 3). Evidence for greater age is 'not conclusive' (Anderson in Hambrey & Harland 1981, p. 765) and, for the fauna, biostratigraphically unacceptable.

An intriguing possibility of a causal nexus between the occurrence of abundant fossils beneath tuff horizons has been discussed at various meetings at which these fossil occurrences were demonstrated. Some or all of the animals may have been killed in periodic volcanic episodes. Most of the fossils appear to represent floating colonial or medusoid Cnidaria but stalked sessile forms also occur. Current ripples suggest transport of benthic and floating elements of the fauna. Detailed taphonomic studies are required of this assemblage which lived and was fossilized under conditions and in sediments different from those at Ediacara, in northern Russia and in the Kuibis Quartzite.

(6) North America. The remaining locality of Late Precambrian Metazoa in eastern North America has yielded so far only one kind of fossil. It has not been found elsewhere. The fossils are worm-like impressions up to 1.1 m long and 15–20 mm wide on bedding surfaces of metatuffs exposed on the banks of a river about 16 km north of Durham, North Carolina. The rocks give a lead–uranium zircon concordia age of $620\pm20$ m.y. The fossils were named *Vermiforma antiqua* Cloud. Meticulous study, comparison and description by Cloud *et al.* (1976) led to the conclusion that the fossils may represent annelids but they could belong to an extinct

phylum. The authors remark that analogy with the Echiura remains plausible but they find it remote. Cloud also considered affinity with Pogonophora and referred to one of Sokolov's (1967) studies of Sabelliditida. He did not give specific grounds for excluding the possibility of *Vermiforma* being related to Pogonophora except the smaller diameter of their tubes (mm instead of cm). However, large Pogonophora have since been discovered living near hydrothermal vents in the deep ocean in the vicinity of the Galapagos Islands; the 'scales' of the fossil tubes may resemble the 'funnels' described by Sokolov as distinctive for certain sabelliditid tubes (Sokolov 1972*a*). The important fact is that there are large, worm-like Metazoa in North Carolina in sediments of known age within the age range of other Ediacarian faunas. Whether the convex 'scales' described by Cloud as spanning the worm tube impressions are related to the funnel structure of some Sabelliditida or whether they correspond to the half-rings formed in annelids (Serpulidae and Cribricyathidae) by deposition from glands under the collar remains to be determined. The relations between Annelida and Pogonophora are still under discussion.

Another recent discovery of a Late Precambrian fossil in North America, the first one from northwestern Canada, was described by Hofmann (1981) as *Inkrylovia* sp. It was found as a float in the area of 'Map Unit 10b' well below the tentative Precambrian–Cambrian boundary (in Map Unit 12) in the west-central Mackenzie Mountains. The name refers to a recently described *Pteridinium*-like leaf-shaped fossil (see p. 87).

(7) Appendix to Section 2.7 – central Iran. A new find of medusae (*Persimedusites chahgazensis*) from the vicinity of Kushk near Bafq, east of Yazd (Fig. 1.8), described by Hahn & Pflug (1980), has drawn attention to a small collection of fossils considered by Stöcklin (1968, p. 1235) as 'closely resembling *Dickinsonia*, *Spriggina*, and *Medusinites asteroides*'. Through the courtesy of the Rio Tinto-Zinc Corporation I was able to examine a 'small but representative collection of fossils' on which this statement was based (letter from A. W. Smith, October 1968). I reported that none of them represented the named Ediacara fossils but that one fragment resembled *Charnia*. This was later mentioned by Stöcklin (1972, p. 204). These finds came from the Rizu Formation of 'Infracambrian' age. The new finds from the equivalent Kushk Shales of the Esfordi Formation made it possible to identify one of the earlier discoveries as *Persimedusites*. It is nearly identical with one of Hahn & Pflug's specimens (1980, Pl.1, Fig. 3). The other fossil fragments from the same 30-cm-thick stratum remain unidentifiable. These authors consider that their new genus

resembles *Rugoconites* Glaessner & Wade. It has also some resemblance with *Hallidaya* Wade and with the 'lobate medusoid' from the Mistaken Point fauna of Newfoundland (Anderson 1978, Fig. 2). According to A.W. Smith the possible *Charnia* was found 'a considerable distance below the black shale horizon', separated by a tectonic break of unknown displacement. Of great interest is the $^{207}$Pb/$^{206}$Pb dating of three samples of ores considered as syngenetic with the black shales. Their ages were 595, 690 and 715 m.y., all $\pm 120$ m.y. (last cited by Hahn & Pflug 1980). A full palaeobiological evaluation of the fauna of the Ediacarian–Lower Cambrian sequence in central and northern Iran will have to await future possibilities of collecting more fossils and stratigraphic and tectonic data.

### 2.8  The concept of the Ediacarian fauna

Having surveyed the Late Precambrian faunal assemblages as far as they are known at the present time, we have to address the question of whether the concept of an Ediacarian fauna is well founded, applicable to all or most of them, and useful in theory and practice. The name itself raises questions but they need not detain us long. It was proposed as early as 1960 by H. & G. Termier in the form of 'L'Ediacarien, premier étage paléontologique'. Since that time many earlier stages in the history of life have been revealed by palaeontological research on microorganisms or on the stromatolitic results of their life activities. Consequently, the Ediacarian is seen as the first palaeozoological rather than palaeontological 'stage' (Cloud & Glaessner 1982). This is not the place for a discussion of wider problems of stratigraphical or geohistorical nomenclature in relation to the Precambrian–Cambrian transition (see Chapter 4). Questions of substance concerning this first known stage in the history of animals have been raised in passing by a number of authors. They can be rephrased as follows: Is the Ediacarian assemblage of fossils a chance association of diverse fossils from diverse places, brought about by specific conditions of palaeoenvironment and preservation at those places, or is it characteristic of a distinctive phase in the evolution and distribution of life in general? Is this assemblage characteristic of a distinctive time interval or is it an arbitrary summation of a number of discrete phases which could and should be separated by more careful analysis of the data?

The first question can be answered by an analysis of a tabulation (Table 2) of the geographic distribution of Late Precambrian Metazoa which summarizes data discussed in detail in the preceding sections and in the literature to which reference was made there. For the present purpose it is useful to list higher taxa, from families to classes and phyla. Questions of taxonomy and phylogeny will be the subject of the next chapter. The tabulation shows clearly that there is in fact a characteristic composition of the Ediacarian fauna. It can be and has been recognized on an almost worldwide scale. In taxonomic terms and in terms of the level of morphological evolution, these faunas are characterized by the relative or absolute abundance of Cnidaria and the considerable diversity of annelid worms and representatives of possibly related phyla. Arthropods are represented by relatively rare and obviously primitive forms. There are a few fossils with no known relatives among either fossil or living animals. From an ecological viewpoint, the animals are benthic (vagile or sessile), nektobenthic, floating or swimming. Some sessile and some floating forms are colonial. Some solitary and some colonial forms tend to reach the maximum sizes of their nearest living relatives. All are characterized by totally or mostly unmineralized tissues but some develop organic or agglutinated tubes. The trace fossils are generally simple, indicating

Table 2. Geographic distribution of metazoan taxa identified in Ediacarian strata ($X_1$ and $X_2$ indicate non-identity of Hydrozoa)

| | | S Australia | C Australia | SW Africa | S America | E Europe | Siberia | China | Britain | Sweden | Newfoundland | N W Canada | N Carolina |
|---|---|---|---|---|---|---|---|---|---|---|---|---|---|
| Coelenterata (Cnidaria) | Hydrozoa | $X_1$ | | | | ? | ? | | | | $X_2$ | | |
| | Scyphozoa | X | X | | | X | X | | | | | | |
| | Conulata | X | | | | ? | | | | | | | |
| | 'Medusoids' | X | X | X | | X | X | | X | X | X | | |
| | Pteridiniidae | X | X | X | | X | X | | | | | X | |
| | Rangeidae | | | X | | | | | | | | | |
| | Charniidae | X | X | | | X | X | X | X | | X | | |
| | Erniettidae | | X | X | | X | X | X | | | | | |
| Annelida (and possibly related phyla) | Cribricyathea | | | | X | ? | | | | | | | |
| | Dickinsoniidae | X | | | | X | | | | | | | |
| | Sprigginidae | X | | ? | | ? | | | | | | | |
| | Anabaritidae | | | | | X | X | | | | | | |
| | Sabelliditida | | | | | X | X | | | | | | ? |
| | Echiura | | | | X | | | | | | | | |
| Trace fossils | Other 'worms' | X | X | X | | X | X | | | | | X | |
| Arthropoda | Vendomiidae | X | | | | X | | | ? | | | | |
| | Parvancorinidae | X | | | | ? | | | | | | | |
| Systematic position unknown | *Redkinia* | | | | | X | | | | | | | |
| | *Tribrachidium* | X | | | | X | | | | | | | |

sediment and detritus feeders or surface grazers. The organisms which produced the traces were mostly worm-like in shape and locomotion but some could have been as complex in morphology and ethology as modern primitive molluscs. There is no indication in this assemblage of special conditions of environment or preservation at the very different localities where these fossils are found. They occur in very different kinds of sedimentary rocks but none indicates origination in deep-water or anoxic conditions. One important factor explaining the abundance of fossilized soft-bodied organisms which are mostly rare in later Phanerozoic assemblages is the absence of macrophagous predators. Incidentally, this characteristic feature of Ediacarian faunas also explains why at many localities these soft-bodied animals could grow to astonishingly large sizes.

The Ediacarian assemblages occur at some distance below fossiliferous basal Cambrian rocks, or in strata which are reliably correlated with those occurring below basal Cambrian in the same region. No mixed Ediacarian and Lower Cambrian assemblages are known. There is no likelihood of the occurrence of remanié Ediacarian fossils in younger strata, as neither trace fossils nor remains of soft-bodied animals can be redeposited. As there was no catastrophic extinction and sudden replacement of the entire biota, single Cambrian taxa (supraspecific at least) can occur in the Ediacarian assemblages and a few Ediacarian lower taxa may have survived into Cambrian time, but the assemblages are distinct, even more than would have been expected. There is no indication of a diachronous boundary in continuous sequences of fossiliferous strata ranging from Precambrian to Cambrian. Such sequences are very hard to find. In most of the likely areas that have been studied, unconformities separate the fossiliferous Upper Precambrian from the Lower Cambrian. These relations will be discussed in some detail in Chapter 4. Here we are concerned with bracketing the Ediacarian between two geohistorical events, one marking its beginning and the other its end (which must be the beginning of the Cambrian), in order to decide whether its subdivision is feasible now or likely to become a practical possibility in the future.

An event which in six major regions predates the first occurrence of Metazoa is the last occurrence of Precambrian tillites (Table 3). Such tillites followed by fossiliferous Precambrian strata are known in Australia, southwest Africa, eastern Europe, Scandinavia, China, Newfoundland, and possibly western North America. In the Flinders Ranges of South Australia the youngest tillite occurs in the Elatina Formation. This is younger than the Tapley Hill Formation ($750\pm53$ m.y.) and older than the Brachina Formation ($676\pm204$ m.y.) This age has an excessive uncertainty but it is confirmed by ages of $670\pm84$ m.y. and $672\pm170$ m.y., obtained from shales above the highest tillites in the Kimberley region of northwestern Australia, the Walsh and Moonlight Valley Tillites (Coats & Preiss 1980). Hence the youngest tillites in Australia are estimated to be 690–700 m.y. old. In southwest Africa the Nama Group rests unconformably on the Numees Tillites which are said to be between 840 and 900 m.y. old (Kröner et al. 1980), but Chumakov (1981) correlates them with a Zairian glaciohorizon (720–750 m.y.). There are reports of glaciogenic deposits in the Schwarzrand Subgroup of the Nama Group (Germs 1972a,

Table 3. *Approximate time sequences of the latest Proterozoic glaciations (triangles), metazoan faunas (circles) and Early Cambrian faunas (squares). Broken lines indicate uncertainty of dating. It involves subjective evaluation of available data and correlations indicating a generalized time framework. Precise correlation within the Lower Cambrian is not implied*

| Approximate scale (m.y.) | Australia | SW Africa | E. Europe | Siberia | China | Britain | Sweden | Newfoundland | N Carolina | |
|---|---|---|---|---|---|---|---|---|---|---|
| 550 | ■ | ■ | ■ | ■ | ■ | ■ | ■ | | | Lower Cambrian |
| 600 | ● | ● | ● | ● | ● | ● | ● | ● | ● | Ediacarian |
| 650 | | | | | | | | | | |
| 700 | ▲ | ▲ | ▲ | | ▲ | | ▲ | | | Varangerian |
| | | | | | | | | | | Riphean |

Kröner & Rankama 1972) but they have been explained as indicating a possibly local cold climate rather than a widespread glaciation. In eastern Europe, the Varangian (or Laplandian) youngest tillites are about 650–670 m.y. old (Chumakov 1978, p. 119). The Nantuo Ice Age in China 'occurred between 740 and 700 m.y.' ago (Hambrey & Harland 1981, p. 944). The tillite in the Conception Group in Newfoundland is dated at about 670–715 m.y.

Ediacarian assemblages comprise almost by definition the first occurrences of metazoan body fossils. It could be thought, and it has been strongly advocated by Cloud on different occasions, that the base of the Ediacarian 'Stage' should be placed at the first appearance of such fossils in stratigraphic sequences. Geochronological and biostratigraphic controls of the various recorded first appearances are still weak and their consideration suggests caution (see above, pp. 92, 98). However these matters are decided, doubts remain about the synchronous appearance of Metazoa in some 15 or more different stratal sequences. I have suggested (Glaessner 1977) that most known Ediacarian metazoan assemblages – which are not likely to include the first Metazoa (see p. 30) – occur in rocks with probable ages between 630–640 and 570–580 m.y. Subject to recalculations based on constants now accepted, and to the fixing of the base of the Cambrian in stratigraphic and geochronometric terms, I see no reason to change this conclusion.

The possible time span of the Ediacarian assemblages, some 60–70 m.y., is comparable to that of the Tertiary, the Cretaceous or the Jurassic, or to the conventional span of the Lower Cambrian including the Tommotian. We can therefore speak of an Ediacarian fauna in the same sense as we speak of the Tertiary fauna as distinct from the Jurassic, or Lower Cambrian fauna. This answers the second question posed at the beginning of this section. The Ediacarian is a definable entity, consisting of different biostratigraphic and biogeographic subunits. This presupposes the possibility of enumerating characteristic features, and this was done for the Ediacarian. It does clearly not indicate complete unity in the sense of indivisibility of this major unit. It is not possible to subdivide it on present knowledge of the nature and fossil content of possible sequential subunits. Assemblages with arthropods or with *Tribrachidium* may well be younger than those without such 'advanced' components but this cannot be proved. Until substantial sequences of fossiliferous strata of Late Precambrian age, particularly in the age range from about 640 to 580 m.y., are found and studied, any attempt at biostratigraphic subdivision or zoning of the Ediacarian must remain speculative and premature. It will be possible in future, given the considerable diversity of marine life which was attained during the first palaeozoologically characterized period of the stratigraphic scale. How this diversity came about will be discussed in the following chapter, while its influence on the scale itself will be considered subsequently.

# 3
# The Precambrian diversification of the Metazoa in the light of palaeozoology

3.1 **The significance of the Ediacarian fauna for metazoan phylogeny**

For the evolution of animals before the appearance of the Ediacarian fauna I have used the term 'Metazoan prehistory' (Section 1.6). Its palaeontological documentation is so poor and so poorly investigated that little can be said about the nature of the animals that existed between about 1000 and 650 m.y. ago. Biological studies of the lower invertebrates tend to give us valuable hints about the kinds of animals likely to have existed at that time. Various more or less probable phylogenies have been suggested. We are beginning to see the reasons why we see few fossil animal remains in the earlier geological record. The phylogenetic results of zoological studies are unbiased because few if any biologists could have known of or would have been interested in Precambrian palaeontological documentation. Since the discoveries of Ediacarian faunas the existence of more or less decipherable remains of Late Precambrian animals has been widely acknowledged in the palaeo- and neobiological literature. There is still a wide gap between the acknowledgement of the existence of Late Precambrian metazoan fossils and their use for the testing of phylogenetic reconstructions or evolutionary hypotheses. Such uses of the newly discovered ancient fossils are increasing but there is still a gap between neo- and palaeobiological interpretations of the record and it seems worthwhile to examine some of the reasons for it.

The traditional view to which many biologists still adhere is that the only palaeontological data worth considering for phylogeny and phylogenetic classification of the living biota are those representing the relatively abundant fossil record of Phanerozoic time in its traditional sense, from the Cambrian onward. Extrapolation into the immediately preceding historical period was commonly based on the expectation that its fossils, when found, would fit onto the known Phanerozoic faunas without a significant break. As many of the existing phyla and classes of invertebrates are represented in the Cambrian, some showing considerable diversity, it is tempting to draw phylogenetic diagrams in which broken lines extend far into the space marked 'Precambrian' until they meet at points of hypothetical common ancestry. On the other hand, doubts about these schemes have led to assumptions of extreme polyphyletism, a separate and parallel evolution of most metazoan phyla from different Protista or even 'eobionts' (Nursall 1962, Fig. 3). Neither of these extreme views has been confirmed by subsequent discoveries of metazoan faunas in the Early Cambrian and Late Precambrian. The expected confirmation of conclusions drawn from the diversity of Cambrian faunas was not forthcoming. It was also thought that because Radiolaria, Foraminifera or sponges are among the most primitive living animals, they should be the first to appear in the Precambrian fossil record. The first reaction to the newly discovered Precambrian faunas was that inasmuch as they did not correspond to these anticipations, they could not be representative of the living fauna of these ages: they must be due to special circumstances of unique ecosystems or of accumulation or preservation of fossils at these times or places. We have seen that there is no basis for these assertions. Another transient idea was that the Lower Cambrian was everywhere transgressive on Upper Precambrian. The implication was that the gaps in the record which in fact exist in many places and are due to a regression of the sea would have obliterated the expected transitional faunas. This was a survival on a minor scale of Walcott's (1910) 'Lipalian interval', a time when sediments with fossils were supposedly deposited only in the depths of oceans while sediments existing now were formed on land. Cloud (1976a) accepted the Ediacarian faunas as representing a time interval preceding the Early Cambrian but drew another conclusion. He regarded it as of sufficiently Phanerozoic aspect, contrasted with the supposedly metazoan-free earlier Precambrian, to be included in the Phanerozoic, representing the latest biostratigraphic division in the history of life which preceded the Cambrian. Questions of major time–stratigraphic classification apart (see Chapter 4), it soon became clear that the earliest Cambrian faunas were as distinctive for their time as the Ediacarian faunas were representative for their earlier stage in the history of animals. The problem posed by the fact that there is no smooth transition from the Late Precambrian to the Early Cambrian fossil record will be examined in the next chapter.

The question now before us is what light the Ediacarian faunas can shed on the diversification of the Metazoa. More precisely, we refer to their particular stage of diversification. Some diversity must have been achieved in earlier times, as shown by preserved though poorly known traces of animal activity. I cannot accept the view that the entire Ediacarian fauna originated very soon after the appearance of the first Eukaryota. It is significantly younger. Eukaryota precede the Late Precambrian glaciations of the period between approximately 800 and 700 m.y. ago, everywhere where relations between these geohistorical events can be established. Ediacarian faunas succeed them (p. 102). No fossils of the Ediacarian assemblage are known from sediments of the Late Precambrian glacial or earlier ages, while Eukaryota are known from sediments which are several hundred million years older. At the present state of our knowledge, we cannot discuss in detail, on the basis of palaeontology alone and in strict sequences, what happened in metazoan diversification from 1000 to 700 m.y. ago, nor establish any *sequence* of metazoan faunas on a sound stratigraphic or geochronological basis for the time interval from 700 to 600 m.y. ago. What we can discuss on a factual basis is the diversity reached in the general Ediacarian assemblage and the biohistorical conclusions to be drawn from it. In particular, we can examine the significance of the apparent high diversity of the Cnidaria and we can assign likely periods of their diversification relative to the broad historical landmark of the Ediacarian Period. We can now estimate the time of the origination of the coelomates and discuss the diversification of the annelids compared with that of the arthropods and with the origination of the lophophorate–echinoderm complex. Most importantly, we can lay the foundations for a comparison, at least in some respects, of the level of evolution of the Metazoa shortly before the beginning of the Cambrian Period with that reached soon afterward, beyond the obvious conclusion drawn in some current literature that the known Early Cambrian animals must have had ancestors. Those comparisons throw light on three much discussed questions: rates of evolution during the early stages of observed global changes of faunas, the validity of some phylogenetic conclusions drawn from biological evidence, and the general dynamics of changes in the biosphere. Therein lies the fundamental significance of the study of Ediacarian faunas.

## 3.2 Re-assessment of the incompleteness of the palaeontological record

It is an undeniable fact that the palaeontological record is incomplete. Only a small percentage of animals living in a given span of time can be preserved as fossils in the sedimentary rocks. The record is grossly biased in respect of size, and of the environment as far as it affects sedimentation and burial of organic remains in sediments. In the preservation of Precambrian animal remains their resistance to destructive diagenetic processes and incipient metamorphism is particularly significant. The record is further biased by selective collecting and description of fossils and by changing aims and interests of palaeontologists and the development of technical methods. At the time when Darwin looked for support of his theory of evolution from fossils, and for one or two generations afterwards, there had been simply not enough collectors and palaeontologists, particularly outside Europe, eastern North America and some colonies of the European nations, for sound conclusions to be based on the nature and distribution of fossils, worldwide. The Precambrian was almost by definition supposed to be unfossiliferous. In recent years the numbers of geologists and palaeontologists discovering, collecting and studying fossils has increased and, in the special field here discussed, the interest in Precambrian studies is growing rapidly, stimulated by the search for mineral resources. It is not surprising that recently several palaeontologists have stated their belief that the fossil record is not nearly as incomplete or as irrelevant for the theory of evolution as has been claimed. To quote Stanley (1979, p.1): 'The role of palaeontology in evolutionary research has been defined narrowly because of a false belief...that the fossil record is woefully incomplete'. He and other authors have proved that conclusions can be drawn from the study of fossils which amplify or correct those drawn from evidence of only the living fauna. In respect of historical events the study of living animals shows at best what could have happened. The fossils supply proof, albeit limited, of what has happened, and when it happened. It can be said that for the solution of many problems the past is the key to the present, without violating the uniformitarian principle of unchanging physicochemical natural laws. The reduction in numbers of animals buried and preserved in sedimentary rocks compared with living faunas of the past loses much of its significance if the bias in nature's sampling process is understood. I have considered the influence of size selection and suitability for preservation when discussing animal Protista (p. 20), mentioned the variety of environments and the influence of factors of sedimentation at various localities where Ediacarian fossils have been found, and have discussed in some detail the constraints on conclusions from the morphology of fossils to that of animals in the fauna preserved at Ediacara. Thorough studies of the taphonomy (burial) and fossilization of soft-bodied animals will supplement the far more advanced knowledge of preservation (and destruction) of shells and skeletons. They are opening possibilites of filling with significant data gaps in the fossil record.

Many examples from the Phanerozoic show that there is no justification for the total neglect of palaeontological data advocated by some zoologists who confine their study of phylogeny to the cross-sectional plane provided by the living fauna. That this 'flat-earth' view is unjustified is shown by much of the information contained in the *Treatise on invertebrate palaeontology* (published by the Geological Society of America and the University of Kansas). Even a single poorly preserved ancient fossil, if properly interpreted, can be worth more than a hypothetically constructed ancestor or archetype, a 'paper animal' as such constructs have been called. A locality with unusually complete preservation such as the mid-Cambrian Burgess Shale (Whittington 1980) can fill gaps in palaeontological knowledge far beyond its own content, by its phylogenetic implications. Newly discovered modes of preservation such as the phosphatized appendages of ostracodes (K. J. Müller 1979, 1981) and new assemblages of small crustaceans from the well known Permian of France and Germany are other examples from the Palaeozoic. Trace fossils give us unexpected insights into the behaviour of extinct animals. A check of numbers of genera at certain points during the adaptive radiation of crabs (Decapoda Brachyura) from their known origination about 185 m.y. ago to the Recent (Glaessner 1969*b*) shows that even among these generally rare and often neglected fossils an exponential growth curve connecting the oldest genus and the point of Recent generic taxonomic diversity (indicating what ideally might be found) does not lie far above the observed points.

The incompleteness of the palaeontological record is not negligible but to some extent assessable, and decreasing. It must not lead to neglect of this source of basic data for the history of the biosphere. That these basic data are no longer being neglected is suggested by a symposium volume, *The origin of major invertebrate groups* (House 1979), in which contributors consider available and relevant palaeontological material.

## 3.3 The record of the Late Precambrian fauna applied to problems of phylogeny

By way of some introduction we note here once again (see also p. 32) the methodological significance of the concept of homology (Remane, 1956) and the similarity between Clark's (1979, pp. 58–9) 'constraints on speculation' and Hanson's 'guidelines' (1977, p. 32; see above, p. 31). Homology is often condemned as the product of circular reasoning and still considered by Clark (1979, p. 57) as theoretically unsatisfactory when it is defined as based on 'structures in existing animals which were transformed from the same structure in a common ancestral animal'. Remane's homology criteria are (i) positional in relation to the fabric of the organism, (ii) qualitative, (iii) existence of transitions in various ontogenetic stages, together with auxiliary criteria of frequency of occurrence, and of exclusion in case of occurrence in unrelated animal taxa.

Clark admits that these criteria 'so far as they can be satisfied, increase the probability that two structures are, indeed, homologous'. Therein lies the significance of homology for the practice of phylogenetic studies which can aim only at increasing probability of correct interpretation of past events. The circularity concerns the logic of the accepted definition of the concept, not its methodological applicability. The alternative to the analysis of the morphology of a fossil organism by application of homology criteria is the observation of unanalysed similarity to other organisms. The probability of correct phylogenetic and taxonomic conclusions on this basis is infinitely smaller. An example is the comparison of the first specimen of *Praecambridium* from Ediacara with the 'similar' probable monoplacophoran *Cambridium*. When more specimens were found, it could be shown that the transverse grooves on the surface of the fossils from Ediacara were not sculptural elements or muscle impressions but segment boundaries, homologous with those in arthropods (Glaessner & Wade 1971). The conclusion to be drawn from the lengthy arguments about the probabilistic concept of homology is that unanalysed similarities of fossil organisms or fossils to extinct or extant ones (or their parts) are frequently misleading and at worst valueless, compared with well-founded and testable homologies. There are numerous examples of phylogenetically meaningless similarities in the living fauna and many are found in recent reviews of the Ediacarian faunas. The following discussion will be based less on arguments against such similarities and more on the specific, stated or implied, homologies.

*Cnidaria.* The first question to be discussed is the adaptive radiation of the Cnidaria which dominate the fauna not only of Ediacara but also of the Nama Group, the Charnian, the Mistaken Point fauna of Newfoundland and the east European Vendian and its equivalents in Siberia. Their worldwide distribution, numbers of individuals and attainment of greater size than at present indicate that the Cnidaria were in fact the dominant element of the macrofauna at that time. This has the advantage for phylogenetic studies of a visible diversity which is open to attempt at interpretation, but it also has disadvantages. Firstly, the observed diversity must have been attained earlier and the paths leading to it are still undocumented. Secondly, the number of finds is so great that many of them are still undescribed; monographic work has not kept pace with discoveries. Thirdly, while there are valid arguments for assigning many of these finds to the coelenterate grade and hence to the Cnidaria, questions of parallel or convergent evolution are undecided and the definition of many species and even genera is still uncertain or controversial, for reasons of fossilization or preservation. Hence, the usual quantitative assessment of diversity by numbers of different taxa in different strata is not yet practicable in an entirely satisfactory manner. Discussing the radiation of the dominant Cnidaria of Ediacarian time, it is necessary to present more technical detail than for other phyla in order to assess the present state of our knowledge and to distinguish facts from speculations.

One of the basic questions in the evolution of the Cnidaria concerns the relations between the extant classes: Hydrozoa, Scyphozoa and Anthozoa. That various *Scyphozoa* were present in the Ediacara fauna was already obvious to its discoverer Sprigg (1947). Dr Mary Wade (in press) has studied recently the distribution in time and the phylogeny of this class (see pp. 54, 114). The presence of *Hydrozoa* can be definitely recognized because of the early appearance of their readily preservable floating forms, the Chondrophora. Other Hydrozoa are believed to be present but are still undescribed. They are recognizable not as medusae, as Sprigg and later interpreters of his work had thought, but as colonial polyps. *Anthozoa* are believed to be present, not as fully mineralized forms but as distinctive frond-like colonies. They are among the assemblage for which Pflug (1972, 1973) has used the names Petalo-organisms or Petalonamae. The morphology and phylogeny of some of them have been restudied by Dr R. J. F. Jenkins but most of his results were not available at the time of writing.

Another basic question which has troubled many students of cnidarian phylogeny is whether the ancestor was a polyp or a medusa. Alternatives to posing this question as a relevant one are either the view that the actinula larva of living cnidarians represents the ancestor, or alternatively that the medusa is no more than a free-swimming inverted polyp or part of one. The significance of this question may be more apparent than real, as medusae can be formed in various ways and at various stages. The event of the appearance of the first cnidarian occurred early in metazoan prehistory, beyond their present fossil record.

The problems of the origin and evolution of the lower Metazoa are discussed, with particular attention to the diversification of the Cnidaria, by Salvini-Plawen (1978). His views on origins differ only in details from those of Ivanov (1968). Their strongly recapitulationist basis may be open to criticism by functional morphologists. These matters of general phylogenetic theory and of details of interpretation are beyond the scope of this discussion. I have accepted the hypothesis of a planuloid ancestry. According to Salvini-Plawen it leads to a bilateral planula which is considered as ancestral to the Bilateria, and to a sessile biradial polyp with nematocysts, the ancestor of all Cnidaria. They, in turn, are divided into the Anthozoa as one clade and all the remaining Cnidaria, all supposedly originally tetraradiate ('Tesserozoa'), as another. Here we see the possibility and necessity of testing a zoological hypothesis against palaeontological data. The view of a derivation of Hydrozoa from Scyphozoa is contradicted by the presence of advanced Hydrozoa (Chondrophora) together with primitive Scyphozoa in the Ediacara fauna. The Chondrophora and the branching colonies of the kind found in Newfoundland are very different Hydrozoa. The former is made up of, from clear evidence of living forms, modified members or evolutionary derivatives of primitive members of the order Athecata. They had been and occasionally still are wrongly associated with the Siphonophora, a separate subclass of the Hydrozoa (Bouillon 1968) which is without fossil representatives. Both the Hydrozoa and the Anthozoa are considered by zoologists to have originated from an actinula-like organism which because of its small size and non-resistant body is unlikely to have been fossilized. It follows that the hydrozoan radiation must have preceded the scyphozoan radiation. Salvini-Plawen's notion (1978, pp. 70, 72) of convergent development of medusae in Scyphozoa and Hydrozoa is acceptable and his statement that 'The phylogenetic position of the Octocorallia is probably the more conservative' compared with that of other Anthozoa is confirmed by the Ediacarian pennatulid-like Charniidae.

The family Charniidae (Glaessner 1979b) has not only close similarities to living Pennatulacea but also distinctly homologous structures (see p. 56). Whether the composite leaf-shaped Pteridiniidae and Rangeidae and the bag-shaped Erniettidae together with the group of *Namalia* Germs and similar genera should be considered as related to the Anthozoa or in part to Hydrozoa or another class remains to be clarified by further studies. There is no doubt that they represent the coelenterate (diploblastic) grade of metazoan organization. As in some other classes of living Cnidaria, there is no trace of a medusa stage related to these organisms.

The fossil record of Early Palaeozoic Anthozoa is increasing, as noted by Scrutton (1979). Of particular interest are the Early Ordovician to Early Devonian fossils described by Bischoff (1978a) as belonging to a new subclass Septodaearia of the Anthozoa. They are described as originally soft-bodied. They are small (about 0.34 mm in diameter), colonial and very variable in septation and other characters. Bischoff is probably correct in placing this subclass on a separate evolutionary line of the Anthozoa. Scrutton's discussion of Early Palaeozoic corals suggests the existence of more such lines at that time.

The place of the Ediacarian fossils in the phylogeny of the Scyphozoa and also their origin and subsequent differentiation are becoming clearer

(Wade in press). Details of theoretical considerations of ontogeny and growth of early Scyphozoa in this work are beyond the scope of the present discussion. The Vendimedusae are considered as the results of an undocumented early or pre-Vendian evolutionary radiation of which one branch, possibly through *Kimberella*, may have led to the Cubomedusida (or Cubozoa), another through the Conchopeltida to the Conulariida. Linked in some way with the origin of this lineage are the 'byroniids', which range from basal Cambrian to Early Devonian (according to Bischoff 1978*b*). They are elongate conical tubes of organic matter, considered as representing the Stephanoscyphus developmental stage of the living Coronatida. Accordingly, a line of descent is assumed to lead from them to the Coronatida, with the later offshoots to the remaining Scyphozoa. While this phylogeny agrees in general with that (Fig. 3.1) proposed by Bischoff (1978*b*), his inclusion of the 'Conodontina' is unacceptable (see Scrutton 1979). Two comments are required. Werner's (1975) study of the entire life history of the cubomedusid *Tripedalia cystophora* is of great importance for the understanding of an unexpected new variant of scyphozoan

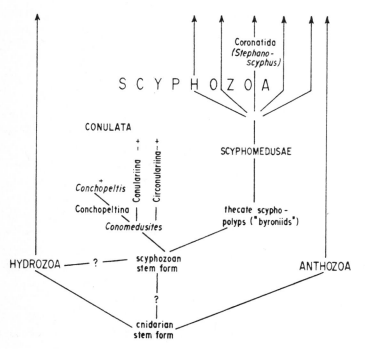

Fig. 3.1. Inferred relations between Hydrozoa, Anthozoa, Scyphozoa, and diversification of early Scyphozoa. Modified from Bischoff (1978*b*).

development. Its polyp stages show budding and can crawl, and finally they undergo a metamorphosis to a tetraradiate medusa. There is no strobilation. Werner's studies led him to establish a class Cubozoa which he placed between Scyphozoa and Hydrozoa. He assumes common ancestors for Cubozoa and Scyphozoa and remarks that Scyphozoa are closer to these (unspecified) ancestors. Dr Wade states that stratigraphic age is against polypoid or medusiform coronates and conulates giving rise to Hydrozoa. Because Cubomedusae have basic scyphozoan characters, it is unlikely that they are either immediate descendants or ancestors of the Hydrozoa. The recognition of a class Cubozoa is not significantly different from the retention of the subclass or order Cubomedusida, as long as this taxon is recognized as closest to the Hydrozoa *among the living Scyphozoa*. Another question concerns the position of the extinct Conulata. I had suggested (Glaessner 1971*b*) that a class Conulata could be derived from Hydrozoa and could give rise, through the Ediacarian to Lower Palaeozoic Conchopeltida and the later *Byronia*, to the class Scyphozoa. Bischoff (1978*b*) studied the septation and recognized what he considered as morphological signs of strobilation in certain Conulata. He placed this taxon as a subclass in the class Scyphozoa. This classification, with its phylogenetic implications, is accepted by Wade who believes that the Vendimedusae, being of the same age as the Ediacarian conchopeltid *Conomedusites* and well differentiated, disprove the conulate origin of the Scyphozoa. These arguments suggest that the Conulata, being less diversified in Ediacarian time than the Vendimedusae, gave rise to the Scyphomedusae through the byroniids but not to all Scyphozoa (Fig. 3.2). This class now includes the Vendimedusae. In this view the Scyphozoa are still conceived as monophyletic but their (unknown) ancestors must have been of pre-Ediacarian age.

These considerations which are necessarily based mainly on fossils from Ediacara, where Cnidaria have been found and described in large numbers and great detail, do not appear to conflict with what is known at present about such fossils from other Ediacarian faunas. The following generalizations appear to be in agreement with the Precambrian fossil record. The earliest cladogenetic event separated Hydrozoa and Anthozoa. Both are capable of developing branching or bunched colonies of polyps but only Hydrozoa developed a floating medusa stage. Early and rapid development of septation characterizing the Anthozoa seems to have precluded the production of a medusa stage. An increasing number of variously septate small corals is being described from Early Palaeozoic strata. They do not fit easily into the standard classification of later and living Anthozoa, suggesting early diversification with much morphological experimentation.

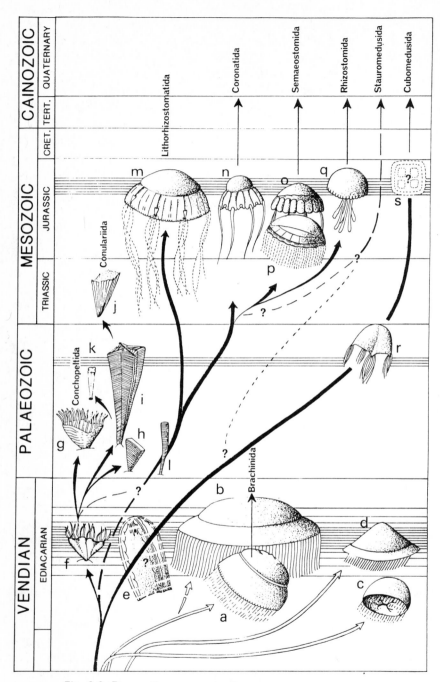

Fig. 3.2. For caption see opposite.

# Late Precambrian fauna and phylogeny

Early polyps were small and could be fossilized only in exceptional circumstances. I have suggested their presence as floating colonies which are common in the Newfoundland fauna and resemble Hydrozoa, and as widely distributed sessile colonies with an attachment disc and a leaf-shaped base. They resemble living Octocorallia (Pennatulacea) in specific characters which have been described as valid homologies. It is impossible, however, to see the detailed characters of the minute polyps which distinguish Hydrozoa from Anthozoa in the living fauna. Both classes could have developed similar benthic colonies, just as the Hydrozoa and Scyphozoa developed 'similar' medusae. One form of colonial organization, that of the chondrophoran Hydrozoa, is complex and provided with a large, preservable chitinous float. There is sufficient evidence to suggest that the Ediacarian fossils assigned to this group do, in fact, represent this taxon. Even the most sceptical commentators have refrained from opposing this suggestion.

The derivation of the abundant and diversified Scyphozoa (three to five genera in the Ediacarian, according to Wade 1983) presents no great difficulty. They do not possess the distinctive characters of the living orders but *Kimberella* may represent the ancestral form of the Cubomedusida. The earliest Conulata (Conchopeltida) may have a similar origin. Before they developed great complexities of their polyp stage, and flourished through Palaeozoic time, the scyphopolyp-like byroniids branched off, leading to the Coronatida and to the other surviving classes.

Many other medusoid fossils in the Ediacarian faunas are at present taxonomically unassignable, in the absence of other than trivial characters of the exumbrellar surface. Some may be hydromedusae, some even parts of Anthozoa, externally similar to but not functioning as medusae. Much work on 'medusoid' fossils remains to be done.

The final question on phylogeny of the Cnidaria is whether there are any indications of other Metazoa evolving from them in Ediacarian time. The

Fig. 3.2. Phylogenetic tree of known fossil scyphozoans.
a – *Brachina delicata* Wade, b – *Ediacaria flindersi* Sprigg,
c – *Hallidaya brueri* Wade, d – *Rugoconites enigmaticus* Glaessner & Wade, e – *Kimberella quadrata* (Glaessner & Wade),
f – *Conomedusites lobatus* Glaessner & Wade, g – *Conchopeltis alternata* Walcott, h – *Conulariella robusta* (Barrande),
i – Conulariina, generalized, j – *Conulariopsis*, generalized,
k – *Circonularia eosilurica* Bischoff, l – *Byronia annulata* Matthew,
m – *Rhizostomites admirandus* Haeckel, n – *Epiphyllina distincta* (Maas), o – *Eulithota fasciculata* Haeckel, p – *Semaeostomites zitteli* Haeckel, q – *Leptobrachites trigonobrachius* Haeckel,
r – *Anthracomedusa turnbulli* Johnson & Richardson,
s – *Quadrimedusa quadrata* (Haeckel). After Wade (in press).

answer is that there is none. This is in complete agreement with most current zoological views. They were sampled by Valentine (1977, Fig. 1) in the form of four selected metazoan phylogenetic schemes proposed in the last 40 years. Only the Ctenophora are considered as derived from Cnidaria, but this is not accepted by Salvini-Plawen (1978). Attempts to link the first finds of Nama fossils with Ctenophora were misguided and are only of historical interest. The extremely insubstantial body of these pelagic coelenterates has precluded their fossilization. The Cnidaria are a successful (11 000 living species) product of early metazoan diversification, without descendants developing in fundamentally new directions. It is probably correct to assume that they have exhausted the evolutionary potential of the diploblastic organisms. Their survival may be due largely to their early acquisition of nematocysts, a superb organ of attack and defence. Further developments, particularly towards more efficient locomotion than the slow creeping of polyps, the passive drifting of colonies or the pulsating swimming of medusae, require a reorganization of the body plan. That this was achieved through neoteny or paedomorphism of a planuloid larva remains an untestable but not improbable hypothesis.

*Coelomates.* The discussions about the origin or origins of the coelom, the secondary body cavity, continued for over a century among zoologists. They were critically reviewed by Clark (1964). His arguments, based on functional morphology rather than on static anatomical or embryological comparisons, are now so widely accepted that no further historical discourse is necessary. Clark's successful functional analysis of problems of metazoan evolution is largely based on histological, physiological and other experimental data derived from living animals. To these data there can be little input from palaeobiology (but see Runnegar 1982*a,b*). An exception to this statement may be made for palaeoichnology, the study of fossil traces of animal activity, which will be briefly considered later (Section 4.8). Our concern here is the testing of phylogenetic conclusions derived from Clark's analysis (Clark 1979). The subject of his review is, as his abstract states explicitly, 'the emergence of different styles of body architecture in animals'. This aspect is at least to some extent testable by comparison with the accumulating evidence of Ediacarian faunas. Because the principal events of the acoelomate radiation are likely to have preceded and those of the lophophorate and deuterostome radiations have, on present knowledge, mostly succeeded the historical period with which we are concerned in this chapter, some of the consequences of the coelomate radiation as presented by Clark will be considered mainly in this context.

Clark's views are stated by him with great caution, appropriate to the complexity and obscurity of the subject. He gives alternative interpretations, either stating reasons for his preference or noting the absence of sufficient evidence for deciding the most likely derivation of a major taxon. He refers to the uncertain definition of a coelom and the probability of an independent origin of its various manifestations in different metazoans. He recognizes the occurrence of evolutionary events at different stages of ontogenesis, including paedomorphism which he links with probable lability of hormonal control of sexual maturation, at an early stage in the evolution of the Bilateria. Finally, the polyphyletic origin of many taxa is admitted but it is not the concern of Clark's work, nor indeed of this chapter, to draw taxonomic conclusions.

These cautionary remarks are necessary to avoid misapprehensions about the accompanying phylogenetic diagram (Fig. 3.3) and the following discussion. Having used Clark's results as a basis, I have made decisions which may appear to be arbitrary, and accepted alternatives about evolutionary pathways which seemed to be more probable on present palaeo- and neobiological evidence. Following Clark I have accepted the sound principle of parsimony ('Occam's Razor'), rejecting more complicated hypothetical pathways where simpler ones are feasible. This is in agreement also with the widely accepted principle of opportunism in evolution (Simpson 1949). Axiomatic approaches to the evolution of the Metazoa have been rejected, such as that of Jägersten (1972), who derived a phylogenetic scheme from the postulate that a pelago-benthonic life cycle is primary for most Metazoa, or the view of many authors that symmetry considerations are basic for the understanding of metazoan body plans. Clark understandably and probably wisely refuses to draw a conclusive diagram of metazoan phylogeny. It seems desirable to construct a model of this process for the purpose of comparing the fauna of a remote historical period with the present fauna and its less distant Phanerozoic predecessors, and to attempt a calibration of the relevant time scale, despite patent gaps in our knowledge.

That the coelomate radiation, discussed at length in its functional aspects by Clark, preceded the Ediacarian faunas can be assumed because of the presence of diversified annelids or at least annelid-grade animals (*Dickinsonia, Spriggina*) and trace fossils of simple ('worm-shaped'), large and hence almost certainly coelomate, infaunal burrowers, and epifaunal thigmotactic grazers. At about the same time there were in existence unsegmented, burrowing coelomates, the Echiura, of which one representative, surprisingly like a living genus, was found in the Nama fauna (Glaessner 1979a). They are often thought to be related to the annelids. Of the Sipunculoidea there is no evidence before the Palaeozoic but this has

Fig. 3.3. Hypothetical stages in the Late Proterozoic and later phylogeny of the higher Metazoa. The sequence of acoelomate and coelomate evolutionary radiations is based on the work of Clark, adapted to show possible origins of major Ediacarian taxa and admitting a possible multiple derivation of the coelom.
(Explanatory notes: 1. Mollusca are here derived from Platyhelminthes. Some Ediacarian and possibly earlier trace fossils may originate from primitive, extinct members of the class Aplacophora. 2. Phylogenetic relations of the arthropods are highly controversial. A possible solution compatible with the primitive Ediacarian arthropods (and with some views of Cisne 1974) is shown. Arthropods are thought to have common ancestors (unknown) with annelids. They may be represented by early trace fossils made by worm-like (coelomate?) metazoans. 3. Original (subsequently reduced or lost) metamerism of a number of living worm-like phyla is controversial. 4. The fossil record of the Phoronida and their origin is uncertain (*Skolithos*?). The Priapulida which are abundant in the Middle Cambrian Burgess Shale are omitted as their coelomate or pseudocoelomate character is uncertain. Other Aschelminthes have no significant fossil record. 5. I have considered the protoconodonts of the Precambrian – Cambrian transition (*Protohertzina*, etc.) as chaetognath spines and at the same time and independently Repetski & Szaniawski (1981) presented a similar interpretation of early 'conodont assemblages'. *Tribachidium* (T) could be related to early echinoderms. The Ediacarian phylogenetic relations of the phyla here grouped as oligomeric coelomates are still unknown. It should be noted that available data indicate early diversification of the 'Spiralia' and do not support the 'Archicoelomate' concept (Ulrich 1972).)

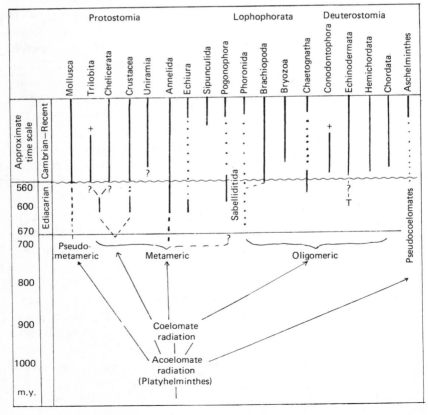

no significance, as their fossil record consists only of one or two more or less characteristic traces of their tube building or feeding. The burrowing Priapulida are well represented in the Middle Cambrian Burgess Shale only (Conway Morris 1977). Some authors include them in the Pseudocoelomata or Aschelminthes of which other constituent taxa include small, sessile, or creeping, often interstitial animals which are only exceptionally fossilized. Clark considers that the pseudocoelomate condition may have evolved more than once, in some clades by paedogenesis, and as part of the early coelomate radiation. Of particular interest in connection with this radiation is the early appearance of probable Pogonophora. Sokolov has insisted in many publications on the placing of the order Sabelliditida Sokolov, 1965 in the Pogonophora. Recent scanning electron microscope studies of the walls of these tubes (Urbanek & Mierzejewska 1977) did not find conclusive evidence for this relationship but did not exclude it. The oldest Sabelliditidae are reported (Wang et al. 1980) from the Sinian of China with *Chuaria* and *Tawuia* in a formation which is probably 800–900 m.y. old (Liulaobei Formation of southern Anhui province), but their similarity with *Tawuia* raises questions about their identification which are at present under investigation. Sabelliditida were also reported from Riphean strata of similar age in Siberia (see p. 90) but according to Sokolov (personal communication, 1979) require confirmation of their age and identification. A convoluted string of fecal pellets about 0.37 mm in diameter in a sediment about 1000 m.y. old in the Riphean of the Ural Mountains (see p. 25) can be reasonably assumed to have been produced by a probably coelomate worm-like animal. If confirmed, the claimed bioturbation structures from the Brioverian of the Channel Islands (Squire 1973), stated to be about 750 m.y. old and measuring 0.5–2 mm in diameter, would agree with the pre-Ediacarian dating of the coelomate radiation and indicate the presence at that time of burrowers and detritus feeders, albeit without disclosing their systematic position. This applies also to a few other reputed small trace fossils of pre-Ediacarian (Riphean) age.

*Arthropoda.* The ultimately most successful product of the coelomate radiation are the arthropods. The Ediacarian faunas contain very different representatives: the family Vendomiidae and the genus *Parvancorina*. The former vaguely resembles trilobite larvae or Merostomata, while *Parvancorina* resembles Marrellomorpha which are remotely similar to trilobites or larvae of branchiopod crustaceans. In the absence of well-preserved legs, which were too thin for the sandy embedding medium in which these fossils occur, no definite conclusions about their systematic placing can be reached except to say that they resemble primitive groups of arthropods in the preserved characters of their bodies. External segmentation as in *Vendia* and development of an external enveloping carapace remain persistent themes in the evolution of the Crustacea. A third kind of primitive arthropod, a chitinous exoskeleton 5 mm wide and 15 mm long, showing many short, undifferentiated segments with pointed pleural margins, was found by Dr M. A. Fedonkin in the Upper Vendian of northern Russia. It is not unlike the Cephalocarida which are now known to have existed in the Late Cambrian (Müller 1981). Only the middle portion is known and appendages are missing in the single specimen from the Vendian (undescribed, personal communication from M. A. Fedonkin 1981).

None of the Ediacarian metameric coelomates indicates the direct derivation of the Arthropoda from Annelida which was assumed in many phylogenetic reconstructions. The positional, structural and functional differences between annelid parapodia and arthropod legs are among the distinguishing features of the two groups. They are in no way bridged by the parapodia of *Spriggina*. The two known genera of the Sprigginidae differ from all known fossil and living annelids, mainly in the configuration of the head. In my first description of *Spriggina* I indicated resemblances with the head of the aberrant, living pelagic annelid *Tomopteris* and their possible significance for the problem of the origin of the arthropods but suggested that more collecting and studying of *Spriggina* should be done before speculating further in this direction. Some results of this work were discussed in Chapter 2 (pp. 61–64). Present views on the evolutionary significance of *Spriggina* are relevant to the consideration of the coelomate radiation, in the light of the continuing discussion concerning this genus (Conway Morris 1979*a*, Birket-Smith 1981*a*,*b*) and the question of arthropod origins (Lauterbach 1973, 1980*a*,*b*). The similarity of the heads of *Tomopteris* and *Spriggina* does not indicate relationship of these otherwise very different and variously adapted genera but only that in a living annelid the head can be laterally expanded although it is generally very small. Acicular setae were present in the parapodia of *Spriggina* which were neither annulated nor articulated. While Manton (1967) and others strongly opposed the possibility of deriving an arthropod limb from a polychaete parapodium, this evolutionary step seems feasible to Lauterbach (1978) and Grasshoff (1981), on the basis of plausible, potentially functional, hypothetical intermediates (Fig. 3.4). However, *Spriggina* shows no specific characters of the arthropods, particularly of trilobites. Its body segmentation and its known appendages are at the level of polychaete annelids but it cannot be considered as a primitive polychaete, having none of the possible ancestral characters indicated in the various conclusions reached by specialists on the systematics and evolution of this group. *Spriggina* should still be considered as an epifaunal, benthic polychaete with a sclerotized head. It represents an evolutionary side line which, metaphorically, could perhaps be considered an unsuccessful attempt to make an arthropod. The arthropods and annelids are both likely to be derived from common ancestors, in the course of the coelomate radiation.

The origin of the Uniramia is uncertain. They are not represented in Ediacarian faunas but appear probably in the Middle Cambrian as *Aysheaia*. This branch of the arthropods which flourished only after it arrived in its dominant, terrestrial habitat, arose separately from other arthropods, though 'both the onychophorans and polychaetes stemmed from burrowing worms with a metameric coelom' (Clark 1979, p. 80). The problem is that there were at the time of the coelomate radiation many different 'burrowing worms' and some of them left trace fossils in rocks of appropriate age, but on that evidence they cannot be distinguished. What was achieved in arthropod evolution by Ediacarian time was a cuticular skeleton, i.e. the development of a flexible cuticle. At least in one instance (*Parvancorina*) it led to a flexible, shield-shaped carapace covering the entire body. It is probable that the lever-type organs of locomotion, the jointed legs on which the designation of the Arthropoda is based, were also developed at that time, but the morphological evidence is unclear. The available data are compatible with a closer connection between Crustacea, Trilobita and Merostomata, as proposed by Cisne (1974) and others.

The great variety of arthropods in the Middle Cambrian Burgess Shale has been commented on by Whittington (1980) who earlier remarked (1979, p. 263) that a major Late Precambrian radiation of arthropods must have occurred 'and that in all probability these arthropods...arose independently many times from metamerically segmented invertebrates (i.e. "worms")'. Of these evolutionary events we have thus far only a faint view from the Burgess Shale horizon. They were a major part of the coelomate radiation in terms of diversity but they proved largely unsuccessful in terms of duration. We see again the opportunistic, multidirectional character of evolution and the unpredictability of the ultimate success of a limited number of body plans and functional adaptations. Lauterbach (1974, 1980a) proposed that the euarthropod appendages developed as a filter-feeding apparatus at the stage of their earliest common ancestor and that the carapace of the Crustacea was uniformly developed from overlapping rearward extensions of the posterolateral margin of the head shield. It is difficult to reconcile these extremely monophyletic views with the limited evidence from the Ediacarian faunas or with the abundant though phylogenetically inconclusive data from the Early and Middle Cambrian.

Fig. 3.4. Hypothetical origin of arthropod legs and skeleton. Diagrammatic cross-sections showing evolution of a thorax segment, from a generalized annelid (top) to a trilobite (bottom). The intermediate stages are hypothetical, implying skeletonization linked to a change in body movements from a predominantly horizontal to a mainly vertical plane, and early evolution of a ventral, filtered, food current. from Grasshoff (1981), where detailed explanations are given; see also Vogel & Gutmann (1981) and compare Lauterbach (1978).

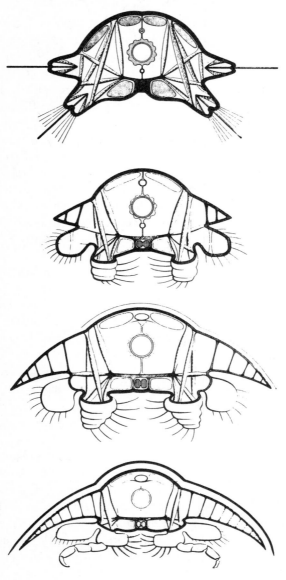

*Mollusca.* The Mollusca were and occasionally still are considered as closely related to the annelids or their immediate ancestors. Clark regards them as pseudocoelomate and pseudometameric and hence 'as part of a radiation of early acoelomate animals' (Clark 1979, p. 67). It is not surprising that molluscan shells are not known from the Precambrian. Palaeontological and phylogenetic concepts of molluscan origins need not begin with shells. Salvini-Plawen (1969) has drawn attention to the phylogenetic significance of the aplacophoran molluscs ('Solenogastres and Caudofoveata'). Together with the Placophora (chitons) they were considered as constituting a subphylum Aculifera, ancestral to the Conchifera. The Aplacophora were subsequently revised by Scheltema (1978) who confirmed the view that they are primitive Mollusca and 'probably very old geologically'. They are worm-like, either peristaltically burrowing (class Caudofoveata = Chaetodermomorpha of Scheltema) or browsing on alcyonarians (class Solenogastres of Salvini-Plawen = Neomeniomorpha = Ventroplicida). The feeding habits of the latter class, described as 'obligate coelenterate feeding' do not appear to have affected the abundant early pennatulacean and similar cnidarians. They may be a later adaptation. The worm-like burrowers among the earliest shell-less molluscs can be expected to be found among the originators of early traces of burrowing Metazoa. Their descendants survived in some abundance in the oceans, including shelves, slopes and the deep sea. Their early representatives would have been less specialized, possibly larger than the millimetre-sized living forms because of less competition. Their spicular integument was possibly homologous to that of their presumed ancestors, the Turbellaria (Rieger & Sterrer 1975). Their descendants may have developed locomotion on a slime trail laid down by a gliding foot. The ventral groove of the Solenogastres is considered its vestige. Some early trace fossils are more complex than 'burrows' made by simple peristaltic movements of undifferentiated 'worms' and may indicate differentiation of a foot and mantle and hence gliding movement of an aplacophoran mollusc. More observations on locomotion trails and excretion products of living Aculifera are required before identification of Precambrian trace fossils as products of this subphylum can be placed on a firmer basis. Analytical locomotion studies rather than comparative configurational iconography will be required before ichnology can date the origin of molluscs and prove the existence of Precambrian aplacophoran Mollusca.

*Oligomeric coelomates and Pseudocoelomates.* The Lophophorata (Bryozoa, Brachiopoda, ?Phoronida) are considered as results of a lophophorate radiation which in parallel with the deuterostome radiation represents the diversification of the trimeric or oligomeric coelomates. There is so far only one animal with a lophophore in the Ediacarian faunas, the still enigmatic *Tribrachidium*, known from Australia and northern Russia. It resembles edrioasteroids which have calcareous plates and are echinoderms. Hence they are deuterostomatous – *Tribrachidium* had no such plates. Ancestors of Brachiopoda which are common in the Early Cambrian would have been small. I have argued against the assumption that the brachiopods must have had shells to protect the lophophore. I thought that a tough, unmineralized, enfolding mantle flap would have sufficed. Now other views have modified and clarified this point. Wright (1979) depicts all diverse brachiopod ancestors as infaunal which means in this context phoronid-like, vermiform lophophorates dwelling in vertical tubes, filter-feeding with their lophophores above the surface (see also Rowell 1981). Gutmann, Vogel & Zorn (1978) derived an analogous model from biomechanical considerations. A strikingly similar model to that of Wright was used by Larwood & Taylor (1979) to illustrate the origin of the Bryozoa. However, their earliest unequivocal representative is from the Lower Ordovician. It has been suggested (Avnimelech 1955, referring to Fenton & Fenton 1934) that some tubular trace fossils were produced by phoronids, particularly *Skolithos*; there are as yet no reliable reports of this fossil from the Precambrian. In Cambrian and Ordovician arenaceous strata these burrows are strikingly parallel and strictly vertically oriented; their builders must have possessed a special organ, probably a statocyst, for their orientation. Their burrows are not likely to have been overlooked if they had occurred in Precambrian quartzites. Other kinds of tubes like those built by living phoronids have been recognized as trace fossils ranging from Devonian to Tertiary. The radiation of the oligomerous coelomates is undocumented in the Precambrian.

According to Clark (1964, p. 219) the pseudocoelomates 'include the Nematoda, Nematomorpha, Kinorhyncha, Gastrotricha, Rotifera, Endoprocta, Acanthocephala and possibly also Priapulida. Some, possibly all, are included in the Aschelminthes.' Most of them are minute and unsuited for fossilization. The Priapulida which are abundantly represented in the Middle Cambrian Burgess Shale (Conway Morris 1977) are now considered by Clark (1979) as non-metameric coelomates, but their ancestry is unknown. Clark (1979, Fig. 9, p. 97) leaves the question of the origin of the pseudocoelomate phyla open. Some may be derived from acoelomates, others from coelomates. A detailed review of the observations on which Clark's views are based is beyond the scope of the present discussion. The same must be said of the observations on which the currently undecided controversy about the affinity of the Pogonophora with metameric coelomates or with deuterostomes is based. New light was thrown on the possible ancestry of at least one branch of the Aschelminthes by the recent publication of an exceptionally well documented hypothesis of the possible derivation of the parasitic Acanthocephala from the mid-Cambrian (Burgess Shale) priapulid *Ancalogon minor* and on the evolution of parasitism in general (Conway Morris & Crompton 1982, see also Conway Morris 1981). These problems are also beyond the scope of the present study.

### 3.4 Occupation of the marine environment: habitats and habits

The Ediacarian faunas provide the first tangible palaeontological evidence of large-scale occupation of the marine environment by Metazoa. I have given a number of reasons why this observed fact cannot be simply translated into the theoretical conclusion that it all happened at once. Clark (1964) has shown that efficient ciliary swimming is limited to animals of small (millimetre) size. Size increase leads to ciliary creeping, with benthic detritus feeding. It precludes sustained burrowing within the sediment until a coelom is developed in a round-bodied, worm-shaped animal with an efficient hydrostatic skeleton. The steps from zooplankton to benthos preceded the coelomate radiation which, as has been shown, must have occurred before Ediacarian time. The results of sustained vagrant burrowing and deposit feeding within the sediment may be observable as bioturbation with backfilling of burrows. Whether this is, as assumed by Clark, a later development than the adaptation to relatively sedentary life in burrows, implying filter feeding, remains to be proved by future sedimentological evidence from Riphean strata and their equivalents. By Ediacarian time, cnidarian polyps and medusae were feeding on organisms within the watermass. The chondrophorans had found ways of exploiting the rich phototrophic plankton of the surface. Nutritional efficiency of cnidarians as well as that of coelomates was sufficient to allow size increase by three orders of magnitude, from millimetre to metre size in individual medusae, in polyp colonies (*Charniodiscus*) and in coelomates (*Dickinsonia* and some tube worms). In the absence of macrophagous predators, large detritus feeders would have flourished, recycling decayed organic nutrients. Nowhere in the Proterozoic stratal sequences do we find an abundance of sediments of an oxygen-deficient marine realm which the Fenchel & Riedl (1970) model of the evolutionary significance of the sulfide biome or the Rhoads & Morse (1971) hypothesis might suggest. Black shales are present throughout the Middle and Late Precambrian but not in greater abundance than in younger strata. Worldwide anoxic phases in the oceans have been suggested for certain Phanerozoic periods in which there was an abundance of marine life, but not for the Proterozoic. If the Platyhelminthes are today primary representatives of the sulfide biome, there is no evidence that they were necessarily its evolutionary product in Precambrian time. It is more likely that they survived in this environment because of reduced predation and competition.

Not all the detritus produced from plankton and plankton consumers was allowed to reach the sea floor and to be accumulated in organic-rich sediments. Some of it was caught by lophophorates (filter feeding). We see little of them in the Ediacarian faunas; they are likely to have remained small until their feeding apparatus could be alternatively raised from or protected by retraction into sclerotized or mineralized tubes or skeletons. Mechanisms similar to those in the oligomeric lophophorates may have operated in metameric sedentary annelids and in ancestral deuterostomes. Speculatively, we may additionally assume population of the sea floor and the infaunal realm by small pseudocoelomates (Aschelminthes), but they appear to be unfossilizable.

It is commonly believed that Porifera (sponges) must have flourished in Precambrian time but in the absence of evidence an alternative hypothesis might be considered. It is possible that they are of later origin. Their evolutionary derivation from choanoflagellate Protozoa need not be dated back to earlier times than the independent origin of the more 'advanced' coelenterates or of the higher invertebrates, the 'Bilateria'. We know that an enormous diversification among the Protozoa, particularly of the Foraminiferida and the Radiolaria, occurred entirely within Phanerozoic time, from very simple beginnings. It seems a survival of the 'ladder of life' idea that Parazoa (Porifera) must have originated after Protozoa, and before Eumetazoa had appeared. The complex and autonomous continuing evolutionary processes among the Protozoa could have reversed, in this instance, the ideally assumed orderly progression. Hanson (1977, Fig. 15.1) also seems to suggest this possibility.

The sea appears to have been populated from the surface to the bottom and eventually to pre-existing or actively excavated spaces within the nutrient-rich sediments, as a result of the efficient hydraulic locomotion apparatus of the coelomates. Absent were the macrophagous predators at the top trophic level, and also the important habitats created later by abundant large brown algae, marine angiosperms and most importantly by algal, sponge, and coral reefs which created in later Phanerozoic times an additional habitat and trophic resource. Stromatolite reefs do not appear to have attracted or supported any fossilizable fauna.

The two available sources of information on the habits of the Ediacarian faunas are structural characters sufficiently close to those of living animals to support conclusions about exploitative and locomotory functions, and trace fossils preserving their effects on the sedimentary substratum or surrounding solid medium. For the abundant Ediacarian Cnidaria, habits of feeding and locomotion similar to those observed in living animals of the same grade of organization can be safely assumed. Polyps as well as medusae extracted food from the biota of the surrounding water and expelled residues and sedimentary particles by contraction. Medusae which were freely deformable could have used the same swimming mechanism as their living representatives (Gutmann 1965). However, for medusoid fossils containing stiff radial rods or other resistant structures which would have impeded such contractions, an alternative interpretation as stiffened, probably chitinous floats as in Chondrophora might be considered. The dorsoventrally flattened annelids *Spriggina*, with traces of aciculae, and *Dickinsonia*, with notopodial or 'notopodial-elytral' (Manton 1967) flexible ridges and with bodies which were flexible to a limited extent in a horizontal (*Spriggina*) or vertical (*Dickinsonia*) plane, could swim or crawl. There is nothing to suggest efficient swimming but only rather intermittent movement from one feeding place on the bottom to another (nektobenthic habit). Close-set meander trails indicate systematic (thigmotactic) grazing of the detritus-covered surface by other worm-like, highly flexible animals which are not preserved. The existence of an echiurid and traces of simple (corophioid) U-shaped burrows indicate semi-sessile life with circulation and exploitation of nutrient-containing water by proboscis activity or peristaltic body movements. Large, smooth trails on the lower surfaces of sandstone slabs suggest sediment feeders able to burrow in clay layers rich in nutrients. Sediment ingestion may be suggested by worm trails cutting across pre-existing moulds of fossils. Preservation of trails may quite generally suggest some cementation of sediment by secreted mucus, which is known to facilitate movement or to be useful for trapping food. Whether originators of some trace fossils were worms or aplacophoran molluscs cannot be decided without further studies. The complex cross-section of the *Didymaulichnus* trail (Fig. 3.5) suggests the gliding movement of a molluscan foot. Another instance of sediment ingestion is provided by pellet trails. Reports of evidence for backfilling of Precambrian burrows are rare, and microscopic studies of Precambrian fossil fecal pellets are inconclusive. Intestinal caeca are known in *Dickinsonia* and suggested in *Praecambridium*. Information on sense organs in Precambrian Metazoa is lacking. It is probable that these organs were of minute size and are unrecognizable in fossil bodies buried or moulded in sand and silt. Chemoreceptors must have been present. Light receptors need not have

130   3: *Palaeontology and Precambrian diversification*

been more than small spots. Antennae of arthropods and similar organs in worms were probably present but too small and fragile to be seen in fossils. It should be remembered, however, that in the absence of macrophagous predators and efficient organs of locomotion required for hunters of prey, complex and efficient eyes would not have had any obvious use in attack or defence in the Ediacarian fauna.

### 3.5 Conclusions on the physical environments of the Late Precambrian Metazoa

Fossil Metazoa occur in various lithofacies of Late Precambrian sediments (see Chapter 2), including shales, siltstones, sandstones and less frequently limestones. Detailed facies analyses of the various fossiliferous sequences have yet to be carried out or completed. Considering the reduced fossilization potential of soft-bodied animals and considering also that it was significantly greater at a time when macrophagous predators and scavengers feeding on the bodies were manifestly absent, we can conclude that in Late Precambrian time at least the continental shelves and epicontinental seas were populated by a variety of Metazoa. Affinities with living invertebrates indicate a normal marine environment. This means normal salinity and absence of significant restriction of water circulation. Palaeotemperatures have not been determined but the composition of the fauna is at least consistent with a temperate to tropical surface and shallow water temperature range. Morphological characters of some body fossils and the occurrence of trace fossils leave no doubt about the existence of a benthic fauna. Its depth range is at present unknown. None of the fossil occurrences which have been described is considered to indicate that the animals lived in deep water. This does not mean that only the shallow seas, shelves and possibly upper slopes were inhabited. Many Late Precambrian deep-water sediments of geosynclinal origin may have been metamorphosed or subducted during Phanerozoic orogenies.

The occurrence of identical species in northern Russia, near the Arctic Circle on the shores of the White Sea, and in South Australia, not far from

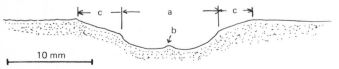

Fig. 3.5. Cross-section of the locomotion trail *Didymaulichnus miettensis* Young. a – double furrow, b – median ridge, c – outer lateral bevels. Top of Miette Group. Rocky Mountains, British Columbia (53° 52' N, 120° 14' W). Modified from Young (1972, Fig. 10).

the Southern Ocean, is a striking indication that the positions of the continents in the Late Precambrian must have been different from those now existing. The movement of marine biotas is least restricted in latitudinal direction at low latitudes. Palaeogeographic reconstructions for the Late Precambrian are still uncertain and are subject to change with continuing accumulation of palaeomagnetic data and their interpretation. It is remarkable that 'conservative' reconstructions, i.e. those leaving continents as close as possible to their present positions, agree with 'radical' solutions in placing the relevant areas close to an equatorial or subtropical position. Examples are Morel & Irving's (1978) map for the 'end of Precambrian, approximately 600 m.y.' and maps by Zonenshain & Gorodnitsky (1977), Ziegler et al. (1979) and Rozanov (1976a) for the Early Cambrian. The wide divergence of these reconstructions shows that we are still far from a sound basis for conclusions on palaeo-oceanography in Late Precambrian and Early Cambrian time. Palaeobiogeography can provide only limited constraints for those reconstructions which permit equatorial or tropical latitudinal pathways.

The unavailability of specific and quantitative palaeogeographic data together with the certainty that conditions were different from those prevailing at present do not allow general conclusions concerning ocean currents, temperatures, salinities and oxygenation of Late Precambrian seas on any firm factual basis. We see no specific reasons in the composition of the fauna for substantial deviations from normal oceanic salinities and warm temperatures. The question of atmospheric and consequent oceanic oxygen availability has been the subject of much speculation and controversy. The Ediacarian faunas show clearly that sufficient oxygen was available for respiration and for the formation of an ozone screen in the upper atmosphere which protected organisms floating on the surface, such as the Chondrophora, from excessive ultraviolet radiation. No difference in oxygen levels in the atmosphere or the oceans from the conditions prevailing at present has to be assumed in explanation of particular characters of the Ediacarian faunas. Sufficient oxygen for respiration and for collagen synthesis (Towe 1970) must have been available when the first Metazoa appeared, several hundred million years before the appearance of the bodies of large Metazoa in the fossil record. I shall return later (Chapter 5) to the question of how much oxygen is sufficient for life activities known to have occurred at particular times.

Measurements of taxonomic diversity play a dominant role in modern ecological and palaeoecological investigations. The diversity of Ediacarian faunas at the highest taxonomic levels (Table 2) is distributed in a peculiar manner. In two areas only, the Flinders Ranges of South Australia and the margins of the East European Platform, each with several localities, this diversity is known to be high. In southwest Africa it is at a similar level. The scarcity of medusae and medusoids in the Lower Nama sediments compared with the abundance of sessile and semi-sessile (burrow-dwelling) animals can be due to different facies or geographic conditions. In the English Midlands and in Newfoundland Cnidaria only are reliably recorded. At the relatively numerous localities where only single finds have been made (zero diversity), they are either medusoid or sessile Cnidaria, with the exception of North Carolina, where only worms have been found. Those occurrences in Siberia which do not appear to have been sufficiently explored for an assessment of faunal diversity or are still essentially undescribed are particularly difficult of access. The diversity of the latest Precambrian or earliest Cambrian there (in the earliest part of the much-discussed Nemakit–Daldyn regional stage) is very small.

When the diversity in the areas with the richest faunas in taxa and individuals is considered at the level of genera and species, we find a striking anomaly compared with later Phanerozoic faunas. From Ediacara 30 species belonging to 21 genera have been described. Fedonkin (1980a) reported from northern Russia 25 species representing 21 genera. Undescribed material which we inspected together in 1979 suggests that the same taxonomic pattern of a high generic and low specific diversity is likely to be maintained when new finds are systematically studied and described. The same applies to undescribed collections from Ediacara. Whether this is an artefact of taxonomy in its application to soft-bodied fossil Metazoa which may give results which are not commensurable with those obtained from fossil shells or skeletons, or whether it indicates a special place of these faunas in the evolutionary diversification of the Metazoa deserves further consideration.

### 3.6 The taxonomy of emerging diversity: the meaning of phyla

According to Simpson (1961, p. 11) 'Taxonomy is the theoretical study of classification, including its bases, principles, procedures and rules'. The principles of classification in a general sense are a vast field of almost impenetrable complexity. Anything passing through our minds, hands or machines can be classified according to any logical principles that serve the diverse purposes for which we classify. The purpose in the present context is the assessment and explanation of the emergent diversity of animals. Avoiding the complexities of the philosophical background it can be said that explanation means here (but admittedly not for everybody) genetic explanation, i.e. elucidation of origination, while diversity requires ordering, i.e. classification. If we were to classify the diversity of a collection of coins or postage stamps, we would base our classification on the fact that they are produced by repetitive actions of relatively simple mechanisms. Classifying the available remains of Precambrian fossil animals which were produced by organic reproduction and growth, interaction with the environment, and fossilization, differs in principle from the classification of mechanically produced objects. In dealing with these, individual variability and preservation may be irrelevant for most purposes, and close similarity of individuals is decisive. Individual variability within populations is a basic assumption in evolutionary classification. Preservation must be considered most carefully because we are concerned with organisms of the past rather than with fossils as objects. The basis of classifying Precambrian animals must be current zoological classification of living animals and their biohistorical predecessors despite the remoteness in time of the processes which produced them. Pragmatic classifications based on chemical (presence of chitin or collagen, or pathways in molecular evolution), ecological or behavioural characters (as determined from trace fossils) are not invalid or excluded but the basic task is ordering within or in relation to zoological classification. Remoteness in time is not an obstacle in principle but a difficulty in practice. Experimental techniques are almost but not totally excluded, as experiments with artificial fossilization or with replication of trace fossils have shown. A large body of biological theory is available to fill gaps in our knowledge of Precambrian animals and can be tested in relation to them.

This is not to deny specific difficulties arising in the classification of early steps in animal diversification. Phylogenetic theory is undeniably in a state of flux, and adherence to one or another school will colour the views of any taxonomist. The fact to be represented in classification is the diversity of objects encountered. In the case of the Ediacarian faunas the diversity of the Cnidaria is in many respects unlike that of their living representatives. Similar observations have been made regarding annelids and arthropods. The unique conditions of the biosphere in Late Precambrian time were such that marine Metazoa, of a sufficiently high evolutionary grade to be free of the constraints of ciliary locomotion and feeding, found many unoccupied ecological niches. Evolution being opportunistic, they had to occupy them, adapt or be preadapted, and diversify. The question is how we classify the known products of this particular diversification event. We classify in a hierarchic system (Simpson 1961, pp. 11ff). At the highest level, most of the Ediacarian animals are at the coelenterate grade, while a minority had reached the coelomate grade. Recognizing that these grades may represent the highest levels in the hierarchy of classification, we refrain from naming this category; we are not concerned here with reforming the system of animals. We avoid the use of terms such as Diploblastica and Triploblastica, Radialia and Bilateria, Spiralia and Archicoelomata or Beklemishev's (1964) superphyla Protrochozoa, Trochozoa and Actinotrochozoa because they are not neutral names for categories but codewords for histological, morphological or embryological contentious theories. Valentine (1973b) re-introduced the term 'superphyla' in a different sense, as a formal taxonomic division of the coelomates, as supposedly natural taxa, adaptively divergent from each other, possessing the general characters of the constituent phyla but without their diagnostic 'modifications and elaborations' and existing during a determinable interval of time (from less than 700–800 to more than 600 m.y. ago). This is dangerously close to a typological (archetypes, Simpson 1961) or a purely operational concept. They are models of adaptive pathways with apparent value for classifying Precambrian organisms, according to the caption of Valentine's Fig. 1. In a sense these models represent the opposite of what appears to have occurred before the Cambrian. The coelomates as well as the coelenterates diversified early into various kinds of organisms. Some of them (e.g. Chondrophora, Vendimedusae, Conchopeltida, *Dickinsonia*, *Protechiurus*) fit into accepted phyla and classes, others fit into a phylum only if its diagnosis is extended (Sprigginidae, Vendomiidae, Pteridiniidae) because they do not fit existing classes; others fit nowhere (*Tribrachidium*). The emerging diversity of metazoans has led to a classificatory dilemma. Either we expand the existing phyla to accommodate taxa which do not fit their present diversity but which appear to be closely related to them, or we adhere precisely to their present definitions and create new phyla. I have followed the first alternative. Phyla were established originally for clearly distinguishable but not necessarily complex and diversified assemblages of living taxa. The number of phyla recognized by different authors varies (15 in Beklemishev 1964, 32 in Margulis 1974, 36 in Barnes 1980), mainly dependent on the status accorded to smaller taxonomic groupings. Most existing phyla with fossilizable representatives date back at least to Palaeozoic times. Few generally recognized phyla are extinct, e.g. Archaeocyatha. The main reason against giving divergent groupings of Precambrian or Cambrian fossils the rank of phyla is the following consideration. We see phyla as the main branches trending upward in a phylogenetic tree, with the vertical axis representing time and the tree seen from above, i.e. from the present. It has often and rightly been said that a three-dimensional phylogenetic diagram would represent a bush rather than a tree because branching occurs early rather than after a number of tall stems have arisen, and many early branches are short. This is what we are now beginning to see as the most probable representation of events in Precambrian–Early

Cambrian time. In early steps in opportunistic evolution, with many new niches being occupied and evolutionary radiations occurring, experimentation takes place, new structural plans evolve, and as a result we find many new clades which proved to be of short duration, for reasons which will be considered. Classification of such short branches gives more useful information when it is related to their origination rather than when the branches are considered as isolated phyla in which the evolutionary experiment of extending diversity in their particular direction has failed. Only those widely diverging side branches of the original metazoan stem or stems are recognized as phyla which maintain their existence for long spans of time and pass through further phases of diversification (major 'speciation events', Stanley 1979). Many of the more recently proposed phyla of marine animals recognized in zoological systematics appear to be of ancient origin, even those which could not be fossilized. Examples are found among representatives of the interstitial fauna of sandy beaches which are small and live in an environment which is hostile to their preservation. It should be noted that initial stages in the emergence of higher taxa could not be recognized as such without knowledge of their further fate: *Archaeopteryx* would have been an aberrant reptile if classified in Jurassic time. In historical perspective, failures are not phyla.

## 3.7 Rates of evolution

Phenotypic evolution can be measured either by morphological or taxonomic criteria (Stanley 1979). Rates of morphological evolution are measured as changes in geological time of sizes or shapes of shells or skeletons or their parts, or gradual increase or decrease of numbers of components. Darwinian evolution occurs in parts of interbreeding populations, i.e. in species, by their isolation and selective alteration of their genetic composition and its phenotypic expression. Opportunities for the study of consecutive stages of species changes within phyletic lineages are rare. Taxonomy expresses differences in terms of morphological characters, producing a hierarchic system of animal diversity. Therefore, taxonomic changes with time which represent rates of diversification lend themselves more easily to quantitative studies. Basically, such rates should be expressed in numbers of sequentially evolving species per unit of time. Stanley has presented much evidence from Phanerozoic history of life in favour of his moderately 'punctuational' viewpoint which asserts that most evolutionary change occurs not within phyletic lineages, as the 'gradualistic' model assumes, but in the branching of lineages ('speciation', an unfortunate choice of a term which etymologically could equally well be applied to phyletic sequences of species). The punctuational model must necessarily be more applicable to evolution in Late Precambrian time when many new lines of Metazoa were being established in an environment which was mostly without pre-existing differentiated Metazoa. Rates of splitting or branching of lineages have to be examined if we are asking whether evolution at that time proceeded at demonstrably higher rates than at other times. We do not know sufficiently extended, strictly consecutive sequences of Precambrian fossiliferous strata and do not have sufficiently precise geochronometry of their scattered occurrences as a basis for their correlation, therefore we have at present no chance of measuring gradual phyletic changes in Precambrian time. We have no chance of choosing between punctuational and gradualistic models on the basis of Precambrian fossils. In the assessment of evolutionary diversification by taxonomic criteria, the species should be the primary unit to be studied, but lack of data often forces the investigator to examine changes with time in numbers of co-existent higher taxa, from genera to orders. Without questioning the value of these data, the best we have if quantitative estimates are to be made, we note an element of arbitrariness which must be well known to any taxonomist who had to classify organisms at family or higher levels. It does not appear to make the data meaningless. What seems more anomalous in the taxonomy of Precambrian faunas is the low number of species in relation to the number of genera, as I have noted (p. 132). It appears that in the early stages of metazoan evolution, major transitions, at least at generic level, were accomplished by quantum evolution (Simpson 1944, 1953), or as Stanley (1979, p. 102) put it, 'by a small number of punctuational steps of great magnitude' effected by means of 'species selection'. This is hardly surprising when the implications of Clark's (1964, 1979) functional explanation of the formation of the coelom and metamerism are considered. Rapid appearance of new genera and of new body plans classified at higher than family level would be expected, and time was insufficient for the phyletic 'fine-tuning', to use Stanley's expression, which would have created numbers of species in and around each lineage. Is there any evidence that these assumed events in early evolution have occurred, and can rates be evaluated? Not only the study but even the collecting and description of Ediacarian faunas is insufficiently advanced to give unequivocal answers to these questions. Stanley has shown that rates of speciation differ depending on habits and habitats, e.g. between bivalves and mammals. In the metameric coelomates of the Ediacara fauna we find many species in the genus *Dickinsonia*, but in the family Sprigginidae it was found necessary to separate the two known species at generic level (Glaessner 1976*b*). Two known families of annelids differed in their mode of locomotion: mainly vertical undulation in

*Dickinsonia* and limited lateral bending assisted by parapodia in the Sprigginidae, suggesting for them a more benthic habit. It appears that in the Ediacara environment the former mode of life was more conducive to diversification at the species level than the latter. Among the Cnidaria, major differences within the floating Chondrophora and the swimming Scyphozoa are expressed only at the generic level, while the numerous individuals of probably sessile benthic medusoids and the definitely sessile colonial Cnidaria are diversified into many genera with probably only one or two species recognizable in most of them. The taxonomic diversity of other Ediacarian faunas, as far as I have been able to study them, seems to exhibit similar characteristic features. There is in this incomplete picture at least a hint that the earlier (Late Precambrian) evolution of most Metazoa proceeded at rates such that almost as soon as new species appeared, they evolved further to produce supraspecific (i.e. generic or family level) differences. This difference between the high rates of evolution at the Precambrian–Phanerozoic transition, admittedly only qualitatively and dimly discerned at the present state of knowledge, and those prevailing at most subsequent times must be related to the unique opportunity of occupation of previously empty ecological niches which followed from the attainment of the metazoan grade, both at the level of the Cnidaria and at the coelomate level. The events of this transition period and their implications for the history of the Metazoa will now be examined in greater detail.

# 4
# The Precambrian–Cambrian transition

4.1 **Stratigraphic scales: boundaries and historical transitions**

For the purposes of this discussion it is necessary to re-examine the stratigraphic classification of the various significant occurrences of Late Proterozoic sedimentary series which are actually or potentially fossiliferous and which range up into Lower Cambrian. Most of the Late Proterozoic sediments of mobile zones must be omitted because of its metamorphism which precludes not only the occurrence of fossils (at least of megafossils) but also the recognition of reliable stratigraphic sequences and determination of their ages of deposition.

Where younger Precambrian sedimentary sequences were recognized on the margins of ancient shields and on stable platforms, attempts to extend the Phanerozoic stratigraphic scale downward have led to proposals of various names for units which originally represented no more than local series of strata and whose new names often lacked strict definitions. Their selective adoption for general use according to historical priority would require extensive and confusing redefinitions in order to make them representative of our present, much expanded, factual knowledge. Accordingly, older names such as Algonkian (rejected in its North American type area but still used in the European literature) and particularly Eocambrian and Infracambrian, with their possible semantic implication of subordination to Cambrian, despite their possible claims to historical priority, will not be used here.

The adoption of a worldwide stratigraphic scale for the Late Precambrian requires definition of its terms by reference to standard sections in which the reality of time-marking events can be demonstrated, and depends on the possibility of worldwide correlation of such events. For different reasons, the application of all methods of worldwide stratigraphic correlation is problematic and can, at best, yield only approximate results.

After much discussion (see Hedberg 1974), prevailing opinion (but see Harrison & Peterman, 1982, for a contrary view) favours the application of a single scale for all geological time, rather than one for the Precambrian and another one for subsequent time intervals and rock sequences. This assimilates the meaning of the marker indicating the beginning of the Cambrian to that of markers of subsequent and preceding major historical intervals or periods. The general, practical, theoretical and nomenclatural problems of geochronological scales are too complex to be discussed here in depth and would lead too far beyond the main subject of this study. Harland (1978) has compiled, after much revising of his and other authors' published work, a clear and comprehensive review of all aspects of the problem. One of them is the dual nature of the divisions of the scale. The boundary between Precambrian and Cambrian seen from the viewpoint of the history of the earth and life is not a sharply defined or definable point in time. No curtain falls and rises again, as between one act and another in a stage play; no lights go on and off again as between two cinema films with different actors. It is, when closely observed, a *transition period*, explicable as a brief speeding up of ongoing processes. In this context, the team 'Precambrian–Cambrian transition' is now used more frequently, and more correctly, than 'boundary'. At the same time, however, an unequivocal boundary line is required to be placed on maps showing the base of the Cambrian. An International Working Group (Cowie 1981) is at present preparing a consensus about the definition of this boundary by selecting a suitable stratotype sequence of rocks in the field and marking (notionally with a 'golden spike') the point where by definition their age changes from Precambrian to Cambrian. This, according to Harland, is a *normative* question which can be settled by a vote. It defines the meaning of the term 'Cambrian' as applicable to all rocks younger than those below the designated boundary (and older than the top of this system, yet to be designated). Which rocks come under this definition and which exposures contain evidence of this boundary becomes a question of stratigraphic correlation, once the boundary is fixed at one or more places. This is an operational boundary, to be placed at some point within the geohistorical transition and to be capable of being extended as a line on a map. The geohistorical transition between two periods of the scale comprises a certain span of time and hence a thickness of rocks deposited during this transition. Compatible with this conceptual distinction is a 'Geochronostratic Scale' (Harland 1978, Fig. 7) which is calibrated with a 'Geochronometric Scale' assigning dates to boundaries and so defining the intervals (see Table 4). This is correct in principle, not just because of the uncertainty of geochronological measurements. When the Subcommission for the Precambrian (Sims 1980) proposed subdividing the Proterozoic at 1600 and 900 m.y., it did not have in mind precise dating which could be determined, with improved methods, to perhaps a million years, but rather concepts like millennia in human history and prehistory (personal communication from H. James, Chairman, 1980). This would divide the Proterozoic into three intervals, respectively and approximately 900, 700 and 330–350 m.y. long, if the Precambrian–Cambrian boundary is placed at 550–570 m.y. Other authors, e.g. Cloud (1976a), explicitly separate geohistoric intervals in their charts by wavy lines spanning 100 m.y. or more in their time scales. Wherever the precise *operational base* of the Cambrian will be placed normatively, presumably by a vote of the International Geological Congress, we are considering here the *geohistorical transition* from the Precambrian to the Early Cambrian. Admittedly, the presently available geochronometric calibration of this transition interval is poor because insufficient dating of relevant rock sequences has so far been carried out.

Before turning to the main subject of this chapter, we must briefly examine another problem of stratigraphic and geohistorical classification. 'The Palaeozoic, Mesozoic and Cenozoic Eras are sometimes grouped together to form the Phanerozoic Eon, as contrasted with another older eon, representing 85% of geologic time, known variously as Precambrian, Cryptozoic, or Archeozoic Eon. The term Precambrian is the most widely used' (International Stratigraphic Guide 1976, p. 78). It has been noted, particularly by Cloud (1976a, and other publications) and Sokolov (1977, etc.) that the meaning of the word Phanerozoic is *visible* (*animal*) *life*, and that if the known history of life is taken into consideration when major divisions of the standard scale are established, the Ediacarian as the first stage containing clearly visible evidence of animal life should be included in the Phanerozoic. I have accepted this argument (Cloud & Glaessner 1982). What becomes then of the widely used term Precambrian? It surely must mean what it says. Cloud proposes the radical solution of rejecting it as a formal term. He replaces the term Proterozoic by the meaningful terms Palaeophytic and Proterophytic, with a boundary at about 2000 m.y. (Cloud 1976a,b), a solution which is unlikely to be recommended for worldwide adoption for the formal time–stratigraphic scale. Cloud does not question the traditional coincidence of the beginning of the Phanerozoic Eon with that of the Palaeozoic Era but Sokolov (1976b, and later papers) does, making the Vendian a pre-Palaeozoic but Phanerozoic division. This makes it possible to continue the use of the established term Precambrian and accords with the status of the Vendian and the Ediacarian as transitional intervals of time at the end of the Precambrian and the

Table 4. 'Geochronostratic' scale of the Precambrian–Cambrian transition. Post-Cambrian period and epoch and post-Early Cambrian age terms are omitted. The terms Vendian, Ediacarian and Varangerian are included as used by Harland & Herod (1975). Some era, period and epoch and the four age terms are selected as examples from Canadian, USSR and Australian regional stratigraphic scales. Their worldwide applicability has yet to be established

| Eon | Era | Period | Epoch | Age | Approximate geochronometric correlation |
|---|---|---|---|---|---|
| Holozoic | Cainozoic | | | | |
| | Mesozoic | | | | |
| | Palaeozoic | Cambrian | Late | | |
| | | | Middle | | |
| | | | Early | Elankian | |
| | | | | Lenian | |
| | | | | Atdabanian | |
| | | | | Tommotian | 550–570 m.y. |
| Precambrian Supereon / Proterozoic | Vendian (Late) | Ediacarian | | | 650–660 m.y. |
| | | Varangerian | Marinoan | | 680–700 m.y. |
| | | | Sturtian | | |
| | Hadrynian | | Torrensian | | |
| | Helikian (Middle) | Riphean Late | | | 950–1000 m.y. |
| | | Middle | | | 1350–1400 m.y. |
| | Aphebian (Early) | Early | | | 1600–1650 m.y. |
| Archaean | | | | | 2500 m.y. |

beginning of the Phanerozoic. The logical extension of the term Phanerozoic to include the transitional final Precambrian time interval leaves the stratigraphic scale without a term for the 'post-Precambrian' Eon. I suggest for it the new term 'Holozoic' and propose that the terms Proterophytic, Palaeophytic and Phanerozoic should be used not as formal parts of the world stratigraphic scale but for divisions of a biohistorical scale. Figuratively speaking, they would be chapter headings in accounts of the history of the biosphere. These eons were characterized, respectively, by the diversification of the Prokaryota, the non-metazoan Eukaryota and the Metazoa.

During the last 30 years the most detailed and comprehensive efforts to establish a Late Precambrian time scale were made in the USSR for the East European and the Siberian Platform, resulting in an agreed stratigraphic scale (Keller 1979, Chumakov & Semikhatov 1981). The Precambrian younger than about 1650 m.y. has been since the 1950s referred to as Riphean, with designated stratotypes in the southwestern Ural Mountains. This term now includes four divisions, from its base to about 680 m.y. The uppermost division of the Proterozoic was designated the Vendian Era or Period. The term Vendian was introduced by Sokolov (1952) and further discussed by him in many papers (1958–1977). It designates Late Precambrian strata of the East European Platform. They rest unconformably on Riphean equivalents and are followed by the earliest fossiliferous Lower Cambrian (Baltic Stage). The placing of the top and the base of the Vendian is still uncertain, both stratigraphically and geochronologically. It comprises, in ascending order, the Vil'chan (or Vilchitsy or Drevlyan) Series with tillites, followed by the Volyn Series with effusives and the Valdai Series. The latter is divided into the Redkino and Kotlin units in the western and central parts of the Russian Platform and their equivalents in the north, east and south (Lower and Upper Valdai Series). These are richly fossiliferous in the north and south, containing equivalents of the Ediacara fauna (see p. 83). The Vendian equivalents on the Siberian Platform are also fossiliferous but in the absence of tillites the base of the Vendian is controversial. It is defined biostratigraphically by stromatolite and microphytolite assemblages.

The isotopic dating of the Late Precambrian of the Russian Platform is extremely unsatisfactory, being based almost entirely on potassium–argon dating of glauconites without sufficient mineralogical studies. This has produced many discordant dates from which stratigraphers have selected evidence supporting divergent opinions. Some geologists wish to confine the Vendian to its upper part, above the glaciogenic strata. The age of the base of the Vendian is accordingly stated variously as $680 \pm 20$ m.y.

(meaning between 660 and 700 m.y.), or 650–680 m.y., or variants of these datings. Nonetheless it appears that the proposed approximate ages of the stratigraphic boundaries in Table 4 are in broad agreement with the latest available datings and acceptable, with possible errors of ±10–20 m.y. These margins were not stated by Harland & Herod (1975) who used the era and period terms for the Precambrian–Cambrian transition as an example in their discussion of the Late Precambrian glaciations. The 'calibration' of this interval is admittedly at present a 'best guess'. Sokolov (1978, p.25) has warned against the use of precise datings of the Precambrian–Cambrian transition, stating: 'The isotopic datings of this boundary, obtained mainly by the K–Ar method, are extremely contradictory and even in the best instances should be accepted with corrections of ±3–5%, i.e. up to 20–30 m.y.'. This becomes highly relevant when we consider quantitative aspects of the increasing faunal diversity which require the placing of faunal assemblages in a calibrated time sequence. The dating of the Early Vendian rests (apart from questionable potassium–argon ages) on the dating of the Scandinavian (Varangian, Varangerian, or 'Laplandian') glaciations. This in turn relies on a rubidium–strontium dating of the Nyborg Shales in Norway which lie between two tillites. It gave an age of $668 \pm 23$ m.y. (Pringle 1973). This adjusts to 654 m.y. when the decay constants agreed internationally in 1976 are used. It can be compared with the dating of two shales in a similar stratigraphic position at about 670 m.y. (Coats & Preiss 1980). These shales lie not far above correlated Late Proterozoic tillites in northwestern Australia. These, in turn, are correlated with the younger one, of Marinoan age, of the two Late Proterozoic tillites in central Australia. The older one corresponds to the Sturtian Tillite in South Australia which is overlain by the Tapley Hill Formation, dated at $750 \pm 53$ m.y. This does not establish a correlation of Late Precambrian glaciations in the Northern and Southern Hemispheres but it places the Varangerian in Scandinavia and the youngest Precambrian glaciations in Australia in the Early Vendian as defined in the stratigraphic scale of the USSR.

The Ediacarian, proposed by H. and G. Termier (1960) as a biostratigraphic term is, on convincing biostratigraphic evidence, an equivalent of the Upper (or Late) Vendian. It has been used as an epoch or period term by Cloud, Harland and others at various times. It was formally proposed as a period term by Jenkins (1981) and explicitly by Cloud & Glaessner (1982) as a term in the world stratigraphic scale. On this occasion an attempt was made to present in graphic form its worldwide correlation, with some of the biostratigraphic evidence for it (Table 5). Details of the suggested correlation are still uncertain. They are hardly relevant for the main subject of the present discussion which is not about stratigraphy as

144    4: *The Precambrian–Cambrian transition*

such but about a framework for the consideration of a significant time of transition in the history of animal life.

The fossiliferous interval following the Vendian in its type area is the Baltic Stage (Sokolov 1952). On the Siberian Platform the carbonates of the Yudomian are followed by either the terrigenous Nemakit–Daldyn 'Horizon' (= Manykay 'Horizon') or by the Archaeocyatha-bearing sediments of the Tommotian Stage (see Sokolov & Khomentovsky 1975). Notwithstanding controversies about its boundaries, the concept and term 'Vendian' is increasingly used outside the USSR where the examination of the biostratigraphic and biohistorical changes at the Vendian–Cambrian boundary began. As others have done, we have to consider the biostratigraphic data obtained there and particularly at the base of the Baltic, Nemakit–Daldyn and Tommotian Stages. Very few Ediacarian fossils have been found in sediments representing these stages. The meaning of this fact for the history of the Precambrian–Cambrian transition will now be examined.

### 4.2    The fate of the Ediacarian faunas: extinction, survival, replacement

In the opinion of some authors, rocks containing families or genera of fossils occurring also in known Cambrian strata could not be of Precambrian (Ediacarian or Vendian) age. Limestones of the Lower Nama Group contain the partly calcareous and partly agglutinated and/or organic worm tubes of the genus *Cloudina* Germs. As this fossil is clearly related to the Cribricyathida which are common in certain Lower Cambrian limestones (Germs 1972a,b, Glaessner 1976a), it was claimed that the entire Nama fauna must be Cambrian. This incorrect view is related to the observation that the base of the Cambrian often seems to be marked by the first appearance of shelly fossils and the erroneous conclusion that therefore none of them could be of Precambrian age. This notion will be further discussed in the following sections of this chapter.

The occurrence of an Ediacarian genus in the Cambrian was equally unexpected. Borovikov (1976) announced a find of a *Dickinsonia* in the trilobite-bearing Lower Cambrian of the Karatau Range, Kasakhstan (USSR). He pointed out the interesting possibility of an occurrence of elements of the Ediacara fauna in younger strata. Having had the opportunity of examining the specimen and discussing it with the author in Leningrad in 1979, I concluded that the figured specimen may indeed be a *Dickinsonia* but because some essential diagnostic characters are obscured by its preservation, confirmation by further finds must be awaited. It is certainly possible that genera and families of the Ediacarian

Table 5. Provisional global correlation of Ediacarian system. (Not to scale; omits many details and some whole regions.) From Cloud & Glaessner (1982). Copyright 1982 by the American Association for the Advancement of Science

faunas survived that period but apparently they amounted only to a small percentage. We must ask what happened to the others.

Phanerozoic soft-bodied organic-walled Metazoa are notoriously rare, at least in our collections if not necessarily in the rocks. It is significant that the few localities where numerous soft-bodied Early Palaeozoic fossils occur (e.g. Lower Cambrian sandstones of southern Sweden, Middle Cambrian Burgess Shale, Ordovician Drabov Quartzite of Bohemia) contain few remains resembling Ediacarian fossils but numerous forms which appear to be endemic to these deposits. The Vendimedusida and other medusoids became extinct without leaving descendants (see Fig. 3.2) but apparently the Ediacarian *Conomedusites* evolved into the Ordovician *Conchopeltis*, with only minor differences. The Chondrophora survived to be represented in the Palaeozoic by at least four genera and Yochelson is describing several additions (see Yochelson & Stanley 1981). They declined later and today they occur as only two species. The Dickinsoniidae may be related to ancestors of the living Spintheridae. The Sabelliditidae survived from Riphean (?) through the Late Vendian to flourish in the Early Cambrian of Eurasia. The living Pogonophora may be their descendants. The Anabaritidae and the conodont-like *Protohertzina* appear in the latest Precambrian and survive in the Early Cambrian. The remaining major taxa, the Vendomiidae and Parvancorinidae among the arthropods and the still enigmatic *Tribrachidium*, appear to be short-lived early offshoots near the evolutionary roots of surviving major clades. At the present state of our knowledge they cannot be confidently claimed to be their ancestors.

Most of the Ediacarian Metazoa died out before the beginning of the Cambrian. How long before the Cambrian? It is not easy to answer this question. Sokolov (1976*a*) refers to the strata containing the Vendian metazoan faunas of the Russian Platform as Redkino Series (Lower Valdai Series), i.e. lower Upper Vendian. He draws attention to the fact that the youngest subdivision of the Vendian, the Kotlin 'Horizon' (or Kanilov Formation) at the top of the Valdai Series contains only trace fossils. It is not possible to determine the vertical range of the metazoan body fossils at Ediacara where one or two unconformities separate the fossiliferous Pre-cambrian from the succeeding Cambrian strata. Sokolov (1976*a*) gives an ecological explanation for his observation on the sequence of assemblages on the Russian Platform. He assumes that heterotrophic consumers including predators, scavengers and bacterial fermenters had increased in numbers before Metazoa had evolved sufficiently resistant chitinous or mineralized exoskeletons which could be preserved. This hypothetical consumer assemblage of Valdai age did not extinguish the Ediacarian biota

## The fate of the Ediacarian faunas

but its destructive trophic activities obliterated it from the fossil record. It is not surprising that this gap in our knowledge of Upper Vendian fossils appears in Sepkoski's tabulations (see Fig. 4.1) of emerging taxonomic diversity which for this interval are based mainly on data from the USSR and Australia. Sepkoski (1979, p. 232) notes that Ediacarian-type fossils largely 'disappear from the fossil record approximately 25 m.y. before the Vendian–Cambrian boundary'. This is more precise than is warranted. Sepkoski holds the accelerating diversification of infaunal organisms and resulting bioturbation responsible for the disappearance of Ediacarian fossils, invoking special taphonomic conditions of the time rather than an evolutionary burst of bacterial activities prior to the beginning of the Cambrian. It seems unlikely that the problem of the fate of the Ediacarian fauna can be clearly defined when we consider that our factual knowledge of faunal sequences is confined to the Russian Platform and not yet

Fig. 4.1. 'Graphs of ordinal diversity in the Vendian and Lower Cambrian plotted against time. The points in the large semilogarithmic graph coincide very closely with a straight line, as expected in exponential diversification. The inset graph illustrates the same data on linear axes; the solid line is a least squares fit of an exponential function ($r = .994$), illustrating the interpreted continuously accelerating diversification across the Precambrian–Cambrian boundary' (Sepkoski 1978, caption to fig. 3). lV – 'Lower Vendian', mV – 'Middle Vendian', uV – 'lower Upper Vendian', ND – 'upper Upper Vendian' (including Nemakit–Daldyn 'Horizon'), T–Tommotian, A–Atdabanian, B – Botomian. Points are plotted at the upper boundaries of these units. Horizontal arrows indicate particular uncertainties of dating. (See Fig. 2 and Appendix in Sepkoski's paper for stratigraphic data, and note criticism of his 'middle' Vendian point in this work, p. 149.)

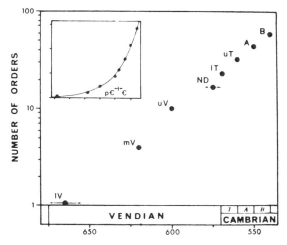

confirmed elsewhere and that the interval in question cannot be dated more precisely than ±10–15 m.y. The taphonomy and trophic ecology of the relevant strata in the region of the Vendian sea discussed by Sokolov are not well enough known to decide between his and Sepkoski's explanation. The fact remains that during Vendian time diversity increased exponentially. Difficulties with its detailed quantitative analysis arise not so much from taphonomic as from taxonomic problems of soft-bodied fossils and from difficulties of placing all finds correctly within a time span of 100 m.y. or less, when intervals of 20 m.y. may be beyond the present resolving power of our dating methods. These difficulties do not preclude acceptance of Sepkoski's general thesis of an exponential increase in taxonomic diversity of Metazoa in Late Precambrian time.

The basis for the quantitative analysis of the emergent diversity during Vendian time will be briefly considered here. The development of the Early Cambrian fauna which involves many questions of taxonomy and phylogeny will be the subject of the next sections. Sepkoski (1978, 1979) bases his analysis on the number of orders of marine invertebrates at three points of Vendian time (680–570 m.y.). They correspond to the upper boundaries of stratigraphic units designated as Lower, Middle and 'lower' Upper Vendian, and a transitional horizon, the 'upper' Upper Vendian or Nemakit–Daldyn Horizon. As both the sequence of emerging taxa and the geochronological position of the data points are based on complex stratigraphic correlations rather than on direct observation of one or more well-dated sequences, their justification must be examined before the conclusion of an incipient exponential, continuously accelerating diversification can be accepted as being in agreement with observed facts.

Sepkoski quotes Lower Vendian trace fossils dated at about 680 ± 20 m.y. as representing one order, Middle Vendian Coelenterata, Conulariida, 'Petalonamae', Annelida *incertae sedis* as four orders dated at 620 m.y., medusoids, Hydroida, Scyphozoa, Conulariida, 'Petalonamae', two or three orders of Annelida, two of Arthropoda, and *Tribrachidium*, as 10 Upper Vendian orders dated at 600 m.y. For the 'upper' Upper Vendian Nemakit–Daldyn 'Horizon' dated at 575 ± 5 m.y., 17–18 orders are listed. Its fauna and age will be discussed below. The following comments relate Sepkoski's assessments to preceding chapters. There are a few rare occurrences of trace fossils which are more than 680 m.y. old (see Chapter 1, Section 1.5 and Glaessner 1983). What order or orders they represent cannot be ascertained but it is immaterial for the conclusion which can only be that diversity at the beginning of Vendian time was very low. The Middle Vendian list is based on Sokolov's (1972*b*, 1973, 1976*a*) lists of fossils from the Mogilev (Podolia) and Lower Yudomian (northern

Siberia) faunas. The fauna of the Yaryshev Formation of Podolia, formerly considered by Sokolov as older, is now considered to be of approximately the same age as the Redkino fauna and this is probably also the age of the fauna with *Glaessnerina sibirica* from the Yudomian of northern Siberia where a similar, rich fauna was collected in 1981 by M. A. Fedonkin and B. S. Sokolov (personal communication). This means that the 'Middle Vendian' data point in Sepkoski's graph (Fig. 4.1) should be fused with that for the 'Upper Vendian' which seems to be based on acceptable data. It is possible or even likely that a gradual increase in diversity during the period from 650 to 600 m.y. will be found when the available material is described, further new discoveries are made, and the geochronology of this interval is refined. In the younger 'transitional' faunas of the Nemakit–Daldyn 'Horizon' of Siberia and its equivalents, more than 10 orders are represented, though not necessarily precisely those quoted from the literature. The significant conclusion is that Sepkoski's graph of exponential growth of taxonomic diversity in terms of metazoan orders, from before 680 to about 580 m.y., represents presently known facts reasonably well and is unlikely to be contradicted by expected adjustments of systematic placement of fossils or dating of strata. It will probably prove to be less smooth than shown in Fig. 4.1. The observed low ratio of species to higher taxa justifies the use of the number of orders for the estimation of the rate of initial diversification of the Metazoa.

## 4.3 The Cambrian Period as the time of the first shelly fossils

Characterizing the Vendian, including the Ediacarian, as a transition period between the Precambrian and the Cambrian leads to the expectation of a smooth transition from the fossil content of the Ediacarian to that of the Cambrian strata. Worldwide investigations of this boundary from a stratigraphic viewpoint and attempts at theoretical evaluation of their results for the history of life have failed to confirm this expectation. This failure is apparent after a wide-ranging discussion about possible explanations for this critical phase in life history. Seventy years ago, Walcott (1910, pp. 14–15) reached an ingenious but erroneous conclusion, proposing the term Lipalian 'for the era of unknown marine sedimentation between the adjustment of pelagic life to littoral conditions and the appearance of the Lower Cambrian fauna. It represents the period between the formation of the Algonkian continents and the earliest encroachment of the Lower Cambrian seas.' Walcott continued: 'The apparently abrupt appearance of the Lower Cambrian fauna is therefore explained by the absence on our present land areas of the sediments, and hence the faunas, of the Lipalian period. This resulted from the continental area being above sea level during the development of the unknown ancestry of the Cambrian fauna.' The historical background to Walcott's statement was elucidated by Yochelson's (1979) thoughtful appreciation of the work of this great pioneer in Precambrian studies. Walcott developed the Lipalian concept in the context of his studies on the American continent which was at that time considered as the type area for Precambrian stratigraphy. It is not surprising that the concept was generalized and applied to other regions where stratigraphic knowledge was grossly inadequate. As knowledge developed worldwide, the concept of a Lipalian interval as an explanation of the sudden appearance of the Cambrian fauna had to be abandoned. Yet, Walcott's definition contains a core of truth. The Lower Cambrian was found in many areas to 'encroach', i.e. to transgress after a widespread though by no means universal regression. As a result of modern detailed studies, Daily (1972) has drawn attention to the widespread transgression of the Lower Cambrian in Australia, Siberia, Scandinavia and elsewhere. The analogy between the regression–transgression events at the beginning of the Cambrian and those at the Permian–Triassic and Cretaceous–Tertiary boundaries on which the recognition of the Palaeozoic, Mesozoic and Cainozoic Eras in the Holozoic Eon rested is striking and its significance for the study of the problem of the sudden appearance of the Cambrian fauna deserves further study.

Another attempt to solve the problem of what happened at the beginning of Cambrian time in a pragmatic manner was propounded mainly by Rozanov (1976*b*) and other workers in the USSR. They contend that the base of the Cambrian is marked by the sudden appearance of 'shelly' fossils, of 'skeletonization' of the previously soft-bodied invertebrates. This view, like Walcott's Lipalian concept, contains a core of truth. The most striking general character of the known Ediacarian faunas is the abundance in them of the remains of soft-bodied organisms, while almost all subsequent Phanerozoic faunas consist mainly of shells and skeletons, with remains of invertebrates without hard parts occurring only rarely.

We must not overlook two important facts. Firstly, a large percentage of living marine invertebrates does not possess hard shells or skeletons. Most of them, or their equally soft-bodied ancestors, are likely to have lived happily without them for hundreds of millions of years. There is little doubt that shells and skeletons are of the greatest significance for geologists and palaeontologists but whether they are as significant for living animals is open to question. Secondly, there is no known gradual transition from the Ediacarian to the Cambrian faunas: the mineralized shells found in the Lower Cambrian are not generally the products of the kinds of animals whose soft bodies are found as Ediacarian fossils. Some authors have proposed palaeoecological explanations of this fact while others have attempted to establish the facts of the Precambrian–Cambrian transition by listing the known Cambrian faunas, at various taxonomic levels, and looking for their possible ancestors or precursors in wide-ranging enquiries about their biohistorical orgins. As recorded in the preceding section (p. 147), Sepkoski has attempted to quantify the apparent suddenness of the appearance of new taxa at the Precambrian–Cambrian boundary. The question of the oldest Cambrian fossils found above Vendian strata where there is no sign of a significant time break is important for the present discussion and therefore the available factual data must be considered in some detail. The meaning of the terms shell and skeleton will be examined in the following section. When that somewhat foggy ground is cleared, the problem of the hypothetical or known ancestry of Early Cambrian Metazoa will be approached.

The greatest exploration efforts concerning the biostratigraphy of the sequence of faunas near the Precambrian–Cambrian boundary have been concentrated on the surroundings of the Siberian Platform, from the shores of the Arctic Ocean to Lake Baikal and from the River Yenisey to within 200 km of the Sea of Okhotsk (Pacific Ocean), an area spanning some 20° of latitude and 50° of longitude. The results obtained in the 1960s were summarized by Missarzhevsky & Rozanov (1968, p. 49) as follows: 'There can be no doubt that two principally different eons exist: the eon of a non-skeletal fauna (Cryptozoic) without any skeletal forms, the eon of a mass development of skeletal faunas (Phanerozoic). The data available enable us to assume the existence between the Crypto- and Phanerozoic of a universal, synchronous boundary associated with a simultaneous appearance (in a geological sense) of a skeleton in most known fossil groups (Spongia, Archaeocyathi, Gastropoda, Hyolithida, Brachiopoda and others), skeletons still being absent in trilobites and some other groups.' The authors established at the base of the Cambrian the Tommotian Stage (Sunnagin and Kenyada Horizons), without trilobites, and claimed that this stage corresponds to strata on the Russian Platform, in England, Morocco and South Australia. 'The strata underlying the deposits of the Tommotian Stage contain, as a rule, only numerous remains of algae...imprints of nonskeletal fossils and tubes of worms.' Details were presented by Rozanov *et al.* in the following year (1969). Further work was carried out subsequently (see contributions by Rozanov, Savitsky, Boudda, G. Choubert, Faure-Muret and others in Cowie & Glaessner 1975); it indicated that the sequence of events at the Precambrian–Cambrian transition and their correlation were not as simple as had been expected. Much attention was concentrated on the strata underlying the deposits of the Tommotian Stage, particularly around the Anabar Shield (or Anteclise) in Arctic Siberia. Bio- and lithofacies and fossil content were different there from those on the Lena and Aldan, 1700 km to the southeast, at the type locality of the Tommotian Stage. It was difficult to establish its equivalents in the clastic facies of the north, named the Nemakit–Daldyn 'Horizon' (regional stage) after one of the rivers draining the Anabar Shield (or Manykay, after another river section further to the east). The problem seems now to have been solved (Zhuravleva *et al.* 1979), in the sense that four pre-Atdabanian fossiliferous zones were recognized in sequences of strata equivalent to or including the Nemakit–Daldyn 'Horizon'. The first one contained only *Anabarites trisulcatus*, *Anabarites tristichus*, *Cambrotubulus* sp., *Sabellidites* sp. and *Protohertzina anabarica* (Fig. 4.2). The second complex of fossils includes the same species,

Fig. 4.2. Latest Precambrian (pre-Tommotian) shelly fossils. A – *Protohertzina anabarica* Missarzhevsky, oblique view and transverse sections. Length about 1.25 mm. Northwest slope of Anabar Massif, northern Siberia. Redrawn from Missarzhevsky (1973). B – *Anabarites trisulcatus* Missarzhevsky, incomplete tube and transverse section. Length about 3 mm. River Yudoma, eastern Siberia. Redrawn from Missarzhevsky, in Rozanov *et al.* (1969).

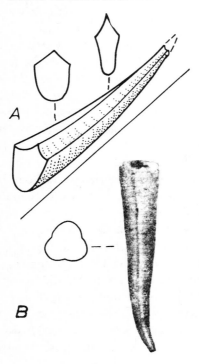

accompanied by hyolithids, *Helcionella*? and in calcareous strata the Cambrian algae *Renalcis*, etc. In the third complex, 20 species including the 'gastropods' *Anabarella*, *Aldanella* and *Bemella* were found and the fourth could be dated as Late Tommotian (Late Kenyada), with the algae *Girvanella* and *Renalcis* being present. An occurrence of Atdabanian Archaeocyatha (fifth complex) was found nearby. The authors indicate that the age of the first complex remains uncertain – the fauna may be Cambrian or latest Precambrian. However, they suggest in conclusion that the base of the Cambrian should be placed at the top of the beds containing the first complex. These beds represent the lower part of the Nemakit–Daldyn 'Horizon'. The upper part (Koril Member) and the following faunal complexes III–IV represent clearly the Lower and Upper Tommotian and the fifth the Atdabanian Stage. It is interesting to note that various parts of this sequence can lie transgressively on older rocks. Only detailed stratigraphic mapping and palaeontological study eventually revealed the complete sequence. There is a transition from the 'first complex' which is considered as Precambrian to the complexes II–IV which can be correlated with the Tommotian and thus are basal Cambrian as presently understood. The algae mentioned by Missarzhevsky & Rozanov (1968) are recognized as Cambrian and placed in complex II, leaving in the Precambrian the fauna with the 'tubes of worms' *Anabarites trisulcatus*, *A. tristichus*, *Cambrotubulus* and *Sabellidites* as 'pre-Tommotian', together with *Protohertzina*.

These somewhat unpromising forerunners of a new era of skeletal organisms herald the appearance of the distinctive Tommotian fauna with its abundant tubular shells and small fossils of unknown affinities. This is biostratigraphically a gradual rather than a sudden appearance of a new kind of faunal assemblage. It does not show any signs of evolving gradually from the preceding Ediacarian fauna. Up to 1981 the only known Ediacarian metazoan fossil from the vast region with Tommotian fossils in northern Siberia was one specimen of *Glaessnerina sibirica* (Sokolov 1972b) from the Khatyspyt Formation, Olenek Uplift, near the mouth of the River Lena. Hundreds of additional Ediacarian fossils were found there in 1981 by B. S. Sokolov and M. A. Fedonkin (personal communication). The lack of transitional faunas linking the Ediacarian with the rapidly increasing numbers of pre-Tommotian and Tommotian taxa is not a local phenomenon. Daily (1972, 1976) has stated that distinctive Tommotian fossils occur in South Australia in lowest Cambrian strata which lie unconformably on the Upper Precambrian including the Pound Quartzite with the Ediacara fauna. Fritz (1980), Conway Morris & Fritz (1980) and Hofmann (1981) reported the discovery of the 'Ediacara-type fossil',

## 4: *The Precambrian–Cambrian transition*

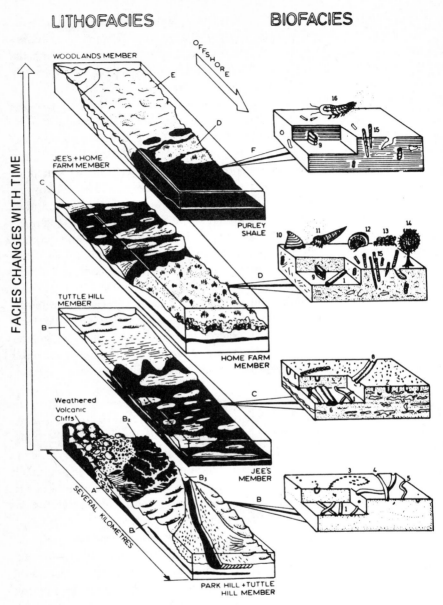

Fig. 4.3. Block diagrams of lithofacies and biofacies in the Hartshill Formation and Purley Shale, illustrating their lateral relationships and changes with time. A, beach facies; B, coastal sands and muds; C, interlaminated sands and muds; D, shelly limestones; E, sheet sands; F, offshore muds. 1, *Planolites*; 2, *Arenicolites*; 3, cf. *Neonereites*; 4, cf. *Psammichnites*; 5, *Gordia*; 6, *Taphrhelminthopsis*; 7, *Isopodichnus*; 8, *Didymaulichnus*;

*Inkrylovia* sp., in Map Unit 10b, and of *Protohertzina* cf. *anabarica* in the overlying Map Unit 11 below the probable base of the Cambrian in the Mackenzie Mountains in northwestern Canada. In 1979 Dr B. Daily and I visited the locality of the occurrence of *Anabarites* above that of an apparent *Charnia*-like frond (*Charnia dengyingensis* Ding & Chen, 1981) in the Dengying Formation of the Yangtze Gorges, below the occurrence of rich Tommotian faunas which were subsequently described in much detail by Chinese authors.

Around the East European Platform and north and northwest of the Baltic Shield we find similar relations between the Vendian and Cambrian faunas. There is an interval of strata of uncertain age which could be assigned to one or the other time–stratigraphic division. The new Cambrian fauna comes in sparingly rather than suddenly and there is no transitional fauna known which would consist of a mixture of taxa from both divisions. Facies conditions were fairly uniform across the boundary. In the northeast (Arkhangelsk) and southwest (River Dniepr) of the platform, the richest known Vendian metazoan faunas have been found in common siltstones and sandstones which in lithological appearance could have been of any age, down to Tertiary. The first Cambrian faunas in the region do not include *Anabarites*, but a *Sabellidites* zone was recognized in Poland, Russia (Rovno Formation) and in northernmost Scandinavia (suggested by Føyn & Glaessner 1979, confirmed by Tynni 1980), followed by a succession of zones with *Platysolenites* and *Aldanella*, *Mobergella* and *Volborthella*, and trilobites. *Platysolenites* does not occur in Siberia; references to it are based on a misinterpretation of a hyolithid (personal communication from A. Rozanov). The genus is also reported from the Lower Cambrian of California.

Tommotian faunas were examined in England by Brasier *et al.* (1978), Brasier (1979) and Brasier & Hewitt (1981) from the Hartshill Formation of Warwickshire (Nuneaton) and from Shropshire. It is separated by tectonic and volcanic events from the Charnian with its Ediacara fauna. The sequence of transgressive Lower Cambrian facies was investigated by Brasier & Hewitt (1979, see also Fig. 4.3). The lowest zone with shelly fossils contains the '*Obolella groomi* fauna' with *Mobergella* (in Shropshire) and *Sunnaginia*, *Hyolithes*, *Hyolithellus*, *Micromitra* and *Coleoloides* at its base and *Camenella*, etc. about 1 m higher. This zone was assigned to the

Caption to Fig. 4.3 (*cont.*)
9, *Teichichnus*; 10, helcionellid mollusc; 11, hyolith; 12, paterinid brachiopod; 13, stromatolites; 14, *Protospongia*; 15, *Coleoloides*; 16, olenellid trilobite. From Brasier & Hewitt (1979). See also Fig. 2.22 (p. 94) for stratigraphy of Hartshill Formation.

*Mobergella* zone of northern and eastern Europe, i.e. to the Late Tommotian (Brasier & Hewitt 1981). Fossils similar to those from England occur in Bornholm and Scania where no Vendian faunas are known, and in Newfoundland and New Brunswick. (A supposed medusoid from the Late Precambrian (Adoudounian) of Morocco (Houzay 1979) is clearly a silicified dolomite concretion in a shale. Apparently no Tommotian shelly fossils have been found in Morocco.)

The facts so far established which require explanations are the apparently 'sudden' disappearance of the Ediacarian faunas, the gradual appearance of *small* tubular fossils, and their rapid diversification in the earliest Cambrian, including shell-like configurations of unknown taxonomic significance, brachiopods, sponge spicules, etc. It was found (p. 146) that at some time between the Ediacarian and Tommotian most of the Ediacarian faunas disappeared from the fossil record. In this somewhat indefinite span of time, of the order of 10–20 m.y., most of the genera and families of the Ediacarian faunas apparently became extinct. However, this observation should not be interpreted as a sudden and dramatic replacement of the soft-bodied Precambrian fauna by the shelly Tommotian assemblages which a literal reading of the record would suggest. Factual observations warn against a simple interpretation. Occurrences of Vendian fossils are rare, on a worldwide scale, compared with Early Cambrian fossils. Metazoa have been found in the Vendian and its equivalents at fewer than 20 localities, despite considerable efforts during the last 30 years. While a few of these localities have yielded thousands of specimens and shown a considerable taxonomic diversity, many others have produced only single finds. In sharp contrast to this, the earliest Cambrian strata are commonly richly fossiliferous although they often yielded large numbers of small fossils of questionable systematic affinities. I suspect that this contrast may be related to an increase in the biomass of heterotrophic organisms. It was possibly based on a similar increase of autotrophs in the sea during the transition from Precambrian to Cambrian. This would be reflected in the overall increase in fossils preserved in the sediments. Apart from biogenic components, Late Precambrian and Cambrian sediments are so similar that taphonomic and other sediment-related factors are unlikely to be responsible for this difference. Admittedly, there are no measurements of probable increases in biomass or estimates of their rates, but the hypothesis may be testable in future by biogeochemical or other methods.

A significant fact to be set against the view that the unprotected soft-bodied Precambrian fauna was replaced suddenly by a fully armed and armoured shelly fauna is the occurrence in the Palaeozoic Era of faunas of soft-bodied marine animals. One example is the well-known Middle

Cambrian Burgess Shale fauna (Conway Morris 1979*b*, Whittington 1980). Another one is dominated, like the Ediacara fauna, by soft-bodied coelenterates. This is the Pennsylvanian (Upper Carboniferous, Westphalian) Mazon Creek fauna from Illinois, where 33% of 3753 specimens of the total biota and 50% of the fauna at the main collecting locality consisted of coelenterates, mainly Scyphozoa (Foster 1979). The fossils occur in siderite concretions in shale deposited in a marine delta. The medusae are assigned to four living scyphozoan orders, extending their known range back from the Upper Jurassic of Solnhofen, where they are found in similar abundance and variety. Corresponding to the similarity with that at Ediacara of dominance of a marine fossil assemblage by cnidarians, is the fact of its preservation in a near-shore environment. The detailed sedimentology and the environmental interpretation of the Pound Quartzite were still unpublished when this chapter was written (see Jenkins *et al.* 1983). It is likely that the environment will be found to be comparable with shifting deltas, but with a sediment supply consisting mainly of quartz sand, with subordinate silt and clay in the vicinity of Ediacara. This example indicates that mass preservation of soft-bodied coelenterates which dominate in a fauna can occur in different taphonomic and environmental circumstances. It could still occur in strata which are only slightly older than 300 m.y. This was to be expected from the abundance of soft-bodied Metazoa in the living marine fauna. I conclude that the Ediacarian faunas were displaced from the fossil record in subsequent eras in the course of the exponential evolutionary diversification of the Metazoa, rather than being simply replaced in the seas by shelly faunas. What were the antecedents of the new arrivals which filled the record so rapidly?

## 4.4 The oldest Cambrian faunas and their evolutionary antecedents

The first appearance of the early Cambrian fauna in great abundance, as observed in northern Siberia and in China, contrasts with its slow and gradual entry in northern Scandinavia, on the East European Platform, and in Australia. Wherever abundant faunas of Tommotian character appear suddenly above sediments of Vendian age without Metazoa, a major disconformity may be assumed to have erased locally some of the record. Alternative explanations are deposition of Precambrian sediments in unfavourable environments, e.g. fluviatile, or there may have been diagenetic alteration such as dolomitization of a limestone. Manifest disconformities may prove, when closely examined, to be caused by submarine erosion and formation of hardgrounds where appropriately specialized faunas lived. A regressive–transgressive phase of sedimentation at the beginning of Cambrian time is generally accepted. Therefore the different kinds of fossil record across the boundary could be considered as a true representation of the biohistorical succession in different places (see e.g. Brasier 1980, 1982). The observed facts on which there is a vastly more abundant documentation than that to which reference can be made here, could represent overall a reasonably complete record, interrupted at different places at slightly different times, but without a major total gap.

The first 'shelly' fossils found in sedimentary strata at the Precambrian–Cambrian transition are not a random selection of a few taxa from the richer Tommotian faunas occurring elsewhere. They are recorded from the Nemakit–Daldyn 'Horizon' by Missarzhevsky, Rozanov, Zhuravleva and other Russian palaeontologists as one or two species each of *Anabarites*, *Cambrotubulus*, *Sabellidites* and *Protohertzina*. The fauna of the lowest zone of the Tommotian (20–30 m higher in one composite section from northern Siberia, Missarzhevsky 1980, Fig. 2) contains 42 taxa and unspecified Archaeocyatha. This is only one example; not all published observations record as striking a contrast between the lowest and the next higher 'shelly' zones. A complete analysis of the fauna of the Lower Tommotian is beyond the scope of the present discussion. It would not be timely because of the large amount of systematic palaeontological work on these faunas now proceeding not only in the USSR but also in China, Scandinavia, England and Wales, Canada, and Australia. Instead, comments will be made on the latest compilations of the earliest Cambrian faunas (Brasier 1979, Fig. 3; Durham 1978, Table 1; Zhuravleva 1970), with due consideration of earlier monographs by Rozanov & Missarzhevsky (1966), Rozanov *et al.* (1969) and later palaeontological work by Matthews & Missarzhevsky (1975), Bengtson (1970, 1976, 1977a, 1978) and others. The aim of this review is to provide a basis of facts from which conclusions will be drawn in the rest of this chapter. Trace fossils will be excluded from the taxonomic part of the review because of the fundamental difference between their taxonomy and that of body fossils. Their significance for the history of life during the transition here considered is not less than but different from that of the 'shelly' body fossils and therefore they will be considered separately.

According to Durham's list (1978, Table 1) the fauna of the lowest (*Aldanocyathus sunnaginicus–Tixitheca licis*) zone of the Tommotian consists of the following groups: Archaeocyatha, Porifera, Hyolitha, gastropods including Monoplacophora, Tommotiida, Hyolithelminthes, Pogonophora, conodonts, and taxa *incertae sedis*. It agrees in general with the list of 'invertebrate macrofossils' given for the (Lower) Tommotian by Brasier (1979) and by Sepkoski (1978). Some comments on this list are required. Bengtson (1977a, p. 51) stated that 'the skeletal fauna occurring at the base of the Tommotian Stage in Siberia represents at least eleven discrete groups of animals with mineralized skeletons'. He includes most of the groups listed by Durham, omitting the Pogonophora which do not have a mineralized skeleton, but not the Hyolithelminthes although their main representative, *Hyolithellus*, is often included in the Pogonophora. The Hyolitha are apparently considered as Mollusca, following Marek & Yochelson (1976). The Brachiopoda are included, as they are by Zhuravleva (1970). The Tommotiida belong to Bengtson's order Mitrosagophora. *Fomitchella* is a conodont in Durham's list but '*incertae sedis*' for Bengtson who lists under family or class names the Sachitidae, Angustiochreida (family Anabaritidae, see Glaessner 1979b) and Coleolidae. Without a complete revision which, as stated above, would not be timely, it is not possible to assign with valid reasons consistent ranks in the systematic hierarchy to these taxa. I have to continue discussing taxa of widely different reported rank, from phyla to genera, without necessarily agreeing with these rankings.

A necessary preliminary task is a clarification of what is meant by the words skeleton or shell in the discussion of the earliest Cambrian fauna. Rozanov (1976b, footnote p. 53) remarks: 'Among the skeletal fossils may be included the anabaritids and sabelliditids which probably are worm tubes. However, any student has the right to call skeletons any hard structures of organisms, from worm tubes even to stone-like cells of plants. Therefore it seems to me absolutely essential to narrow down this concept. At any rate it should be emphasized that in Early Tommotian time we meet for the first time mollusc-type skeletons and a number of other specific types. The principal difference between skeletons of this type and the tubes (or dwelling structures) of worms and possibly related groups is of course clear.' Of course it is not clear, otherwise less effort would have been spent on means of distinguishing, for example, between tubes of gastropods, scaphopods and annelid worms. It has become common practice to refer to *skeletons*, complex solid structures which may be linked or fused, the latter kind occurring in Archaeocyatha; *spicules* (isolated or fused elements, small in relation to the whole organisms), e.g. in Porifera; *shells* in brachiopods, bivalved or segmented mineralized arthropod integuments, and in molluscs and hyolithids (where they may be tubular); to *tubes* in *Hyolithellus*, *Sabellidites*, and other worm-like organisms. These tubes may be organic, phosphatic or calcareous, or they may be wholly or partly agglutinated from sedimentary particles with mucous or calcareous cement. This leaves one group, the separate, multiple, mineralized elements of irregular or varied shape. They may have been numerous but not firmly linked in the integuments of animal bodies. The term *sclerites* is commonly used for them. Bengtson & Missarzhevsky (1981) have ascribed three different kinds of *hollow* sclerites to a class Coeloscleritophora which they assumed to have been monophyletic. This seems to be a far-reaching conclusion based on a single common character. Bengtson (1977*a*) noted that tubes of *Anabarites, Sabellidites* (or *Paleolina*), the winding tubes of *Cambrotubulus* and the sclerites of *Protohertzina* occur in pre-Tommotian strata and that therefore it is not correct to say that all mineralized skeletons appeared simultaneously. The Tommotian 'skeletal' fauna appeared in a process that 'did not take more than a few million years at the most'. Before considering why this mass appearance was apparently so hurried, we must ask with Durham (1978), 'from what ancestors or precursors did this fauna originate?'.

The problem of the origin of the rich Tommotian fauna can be broken down into several different questions regarding different taxa or groups of them which are known to have originated at different times or in different ways. The first group which may date back to pre-Vendian (Riphean) times are the Sabelliditida which are believed to be the oldest Pogonophora. Zoologists are becoming increasingly inclined to link these aberrant worms with the Annelida which may also date back to the Late Riphean. Poulsen (1963) has linked *Hyolithellus* with the Pogonophora, on the basis of supposed chemical similarities of their tubes. The analytical data may not be fully convincing but most palaeontologists seem to have accepted Poulsen's arguments. Assignment of this group (probably including *Torellella*) to the Pogonophora may remove its origination from the question of the sudden origination of new elements of the Tommotian fauna. A second group includes the Archaeocyatha whose phylogeny has been studied in sufficient detail to give a coherent picture of their history from beginning to end. According to Rozanov (1973) and Debrenne & James (1981) (see Fig. 4.4) they commenced with a small number of genera and species at the beginning of the Tommotian and disappeared at or near the end of the Lower Cambrian. Their genetic relations to other taxa is unknown. It is thought that there are similarities between their origination and that of the Porifera from protozoan colonies, but no morphological homologies can be found. Öpik's (1975) suggestion that they are plants was not supported by firm morphological data and has not been widely accepted by specialists. The question of a connection with green algae (receptaculitids) was re-opened by Nitecki *et al.* (1981). The Archaeocyatha could have originated from unknown small colonial protists, just before the Cambrian. The important fact is that they were able to build calcareous skeletons with gradual adaptations facilitating the flow of nutrient-bearing water through the intervallum. There are no reasons why they should be considered either as coelomates or as derived from acoelomate Metazoa.

The Porifera which appear at the beginning of the Cambrian (reports from earlier strata having proved unreliable) show a similar picture of limited diversity among the oldest representatives, with a reasonably coherent record of subsequent evolution. They did not share the fate of the Archaeocyatha. At the beginning of the Cambrian, mineralized spicules appear in low taxonomic diversity but abundantly.

The 'phylum' Cribricyathi has caused much confusion. Although these fossils were first described as Cambrian Archaeocyatha, the morphology of their tubes is entirely unrelated to the characters of that phylum. Germs (1972b) discovered them in limestones at various levels in the Lower Nama Group and recognized their affinity with annelid tubes which I have confirmed (Glaessner 1976a). Annelids range probably from Riphean and definitely from Ediacarian to the present. There are many other kinds of thin and long, generally small, tubular fossils among the Early Cambrian faunas. What is known about many of these thin-shelled tubes does not

Fig. 4.4. The number of archaeocyathan genera plotted against time for the Lower Cambrian. Stages at left: (1) North American trilobite zonation, (2) Lower Cambrian stages of the Siberian Platform, USSR. (Solo. – Solontzy Horizon, Obr. – Obruchev Horizon). After Debrenne & James (1981).

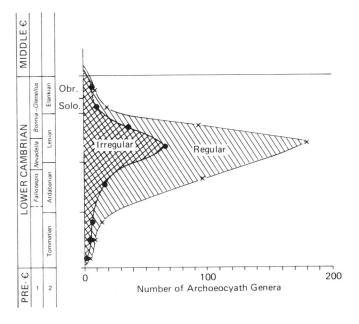

enable us to distinguish between metameric annelids and oligomeric or unsegmented coelomates as their producers. Diagnostic characters may yet be found in the microstructure of their tubes. I have suggested that the tubes of the Angustiochreida (Family Anabaritidae) may have been produced by annelids because of their ornamentation which has significant similarities with that of serpulid worm tubes. The Coleolidae form thin, long tubes of unknown affinity. Some of them seem to have been flexible when buried in the sediment, suggesting partly organic composition. There are no grounds for considering them as planktonic Mollusca. There is still little known about Brachiopoda from the basal Cambrian of the USSR or Australia. Two Acrotretidae from the Aldan River were figured by Zhuravleva (1970). Four orders (Lingulida, Paterinida, Kutorginida and Discinida) are listed as occurring in the early Lower Cambrian by Wright (1979). It is possible that some of the early brachiopods secreted chitino-phosphatic tubes which may lack distinctive characters and therefore appear in the fossil record of the early Tommotian as *incertae sedis*.

The sudden appearance of the Early Cambrian Metazoa which were considered thus far, present no major problem regarding their ancestry. Either they have known roots in the more distant past (Sabelliditidae, Cribricyatha, annelids in general) or they show the beginnings of a complex subsequent history of diversification (Archaeocyatha, Brachiopoda) from very small, soft-bodied or soft-shelled, unknown but for valid reasons acceptable, hypothetical ancestors. This leaves some more problematical groups: the conodont-like sclerites which appear with the first tubular fossils just before the Tommotian, and the hyolithids which are found in great abundance, often together with other more or less molluscan-like 'shells' which appear at the beginning of the Cambrian, the Mitrosagophora.

The conodont-like fossils from the Tommotian are elongate, curved, claw-like objects, 1.2–2.75 mm long, with a convex side and a sharp keel between two concave zones on the opposite side. These vaguely toothlike objects were named *Protohertzina* by Missarzhevsky (1973) who compared them with the proboscis spines of the priapulid worms. This suggestion was generally disregarded because he placed *Protohertzina* in the conodont family Distacodontidae. It deserves reconsideration because priapulids have since been recognized as abundant fossils in the Cambrian Burgess Shale (Conway Morris 1977). *Protohertzina* resembles closely the grasping spines of the oligomerous coelomate Chaetognatha which are considered by Hyman (1959, p. 226) as possibly early offshoots of the main deuterostomatous line. This may bring them in the vicinity of lophophorate origins. With reference to later Cambrian and Early Ordovician conodont-like fossils, the homology with grasping spines of chaetognaths was independently proposed by Repetski & Szaniawski (1981; see also Szaniawski 1982). Lophophore supports of a Middle Cambrian fossil (*Odontogriphus* Conway Morris, 1976) have also been compared with Cambrian conodonts. One kind of these comprises the funnel-shaped spiny objects named *Fomitchella* Missarzhevsky, 1969. However, Bengtson (1977a) found their microstructure incompatible with that of the other Cambrian conodont-like forms. Some of these he separated as protoconodonts, which are histologically external elements of the body, as are the priapulid and chaetognath spines. Bengtson admits that there is no direct evidence for the derivation of paraconodonts from protoconodonts, which he considers as a separate lineage throughout the Late Cambrian. The euconodont animal, the bearer of elaborate assemblages of diverse elements unlike any lophophore, is still unknown but it was probably unrelated to *Protohertzina*.

The Hyolitha are considered by Runnegar *et al.* (1975; see also Runnegar & Jell 1976, and Runnegar 1980a) as a phylum separate from but closely related to the Mollusca, but Marek & Yochelson (1976) see no reason 'at this time' for excluding them from the mollusc phylum. These authors admit that Hyolitha do not fit into some schemes of molluscan phylogeny but they do not propose one into which their elongate conical operculate tubes will fit. It is difficult to see for what functional advantage in response to selection pressure ('evolutionary rationale', Stasek 1972) some primitive Mollusca would have formed elongate conical tubes. Most zoologists now accept the turbellarian origins of the Mollusca. Stasek's reconstruction of stages leading through a stage of a covering cuticle to a calcified shell has been confirmed by the study of numerous Tommotian Monoplacophora and their descendants (Runnegar & Jell 1976, Pojeta 1980). Compared with them, the Hyolitha must have been adapted to a very different, less mobile mode of life. It is probable that their ancestor was a semi-sessile worm, different from the creeping progenitor of the molluscs.

A large complex of sclerites was grouped by Bengtson (1970, 1977a, 1978) as an order Mitrosagophora, hypothetically linked with the later Machaeridia. Much light was thrown on this taxon by the discussion of a new find of *Plumulites* Barrande (Early Ordovician to Early Devonian) by Jell (1979), who concluded that the taxa Machaeridia and Mitrosagophora may eventually become synonymous. Both terms refer to bilaterally symmetrical, metameric or, more likely, pseudometameric animals covered dorsally with imbricate sclerites of calcareous or phosphatic composition. Mitrosagophoran assemblages of different sclerites from single animals have not yet been found but I have concluded (from unpublished observations) that they must have existed at least in the later part of the Early Cambrian (Botomian) *Tannuolina*. I cannot agree with Jell's conclusion that the group discussed by him probably belonged to the phylum Annelida. Their elytra are very different in form, growth and function from sclerites of *Plumulites*. This was confirmed recently by Conway Morris (1982) who is re-investigating the mid-Cambrian Burgess Shale fossil *Wiwaxia* Walcott. He found also 'that the similarity between the aplacophoran spicules and the [wiwaxiid] sclerites is superficial', and that similarities tentatively considered by Jell (1981) with opisthobranch cerata are 'entirely superficial'. On the other hand he considered a relationship between the Wiwaxiidae and the Early Cambrian phosphatic sclerites *Sachites*, *Halkieria* and others (collectively named Thambetolepida Jell) 'plausible' but an affinity with the Mollusca 'remote'. The various observations made by Bengtson and others, and the lack of any tendency towards articulation of the sclerites in Mitrosagophora, removes them from the Polyplacophora. On the other hand, the Aplacophora and their hypothetical ancestry (p. 125) deserve consideration and may provide clues to their systematic position. Their morphological links with early Mollusca are significant. Their organization is primitive in many respects but they differ from the ancestors of the other Mollusca in the reduction of the distinctive organ of locomotion of that phylum, the foot.

On the basis of these observations, which should be supplemented by much further research, a working hypothesis about the evolutionary origins of the extinct Mitrosagophora–Machaeridia complex and possibly other Early Cambrian sclerite-bearing taxa (*Thambetolepis* Jell, *Sachites* Meshkova, *Halkieria* Poulsen) may be proposed. It is probable that these bilaterally symmetrical animals which were covered with columns of at least two different kinds of sclerites evolved from ancestors of shell-less Mollusca (Aculifera). They had a cuticle which developed calcareous sclerites varying in shape from spicular to conical, pyramidal or scale-like, but not articulating with each other and without any tendency for the sclerites to fuse into a single plate or a small number of similar transverse plates. Their scaly dorsal covering excluded the possibility of effective peristaltic movement. Locomotion could only be effected by means of a ventral foot. Organs of respiration had to be located laterally, in a mantle fold between the foot and the dorsal armour. Other parallel evolutionary tendencies to those of the Mollusca would have been present if they all possessed the same three basic systems: foot, mantle and organs of respiration. The non-spicular sclerites did not develop or were lost in the Aculifera, which showed a tendency to enroll permanently in a transverse direction (Stasek 1972, Fig. 3), assuming a round worm shape, as did some

Machaeridia (Bengtson 1970, p. 387). In the Aculifera this tendency led to a restriction of the foot to a median ventral strip (Ventroplicida).

The numerous common ancestral and the few convergent derived characters do not suggest that the mitrosagophorans are molluscs. Given the totally different paths taken in the evolution of the main molluscan classes (Runnegar & Pojeta 1974; a dissenting view on their phylogeny is presented by Yochelson 1978), the tentative conclusion is that the complex of Mitrosagophora and Machaeridia evolved in the Early Cambrian from common ancestors with the Mollusca, in a fundamentally different direction from that of the now dominant representatives of this phylum, and became extinct at the end of the Palaeozoic Era. The erroneous views of echinoderm or arthropod (cirriped) affinities of the Machaeridia have been exhaustively discussed and dismissed by Bengtson (1970, 1977a, 1978). Annelid affinities cannot be accepted in the absence of evidence of true metamerism, of parapodia, the unlikely homology of machaeridian sclerites with elytra, and the different developmental origin of mineralization of annelid tubes.

The discussion of the oldest Cambrian faunas and their ancestry cannot be concluded without reference to the Arthropoda and particularly the trilobites. Arthropoda occur in the Ediacarian faunas. Their preserved remains do not present details of their locomotry organs which are essential for their more precise classification. What is known indicates their resemblance to very primitive representatives of this phylum. Rozanov and other stratigraphers in the USSR have indicated that the Tommotian Stage is characterized by the absence of trilobites. Before this view was challenged by a reported find of a trilobite in the stratotypic Tommotian (Fyodorov, Yegorova & Savitsky 1979), it had been questioned for two reasons. Firstly, many critics consider it unlikely that the considerable diversity of post-Tommotian (Atdabanian) trilobites could have originated simultaneously, and secondly, it is agreed that Tommotian trace fossils resemble closely some (*Rusophycus*) that were made demonstrably by trilobites. In Morocco, trilobites were found in Early Cambrian strata but they have not yet been fully described, and restricted Tommotian fossils have not yet been found there. The presence of *Rusophycus* can be interpreted to mean that arthropods with sclerotized legs existed and behaved like trilobites (but so do some notostracan crustaceans and some merostomes). It is reasonable to assume that early representatives of trilobites existed during Tommotian time but most of them were not preserved because they were either too small, or unsclerotized, or unmineralized. The application of the concept of the Tommotian Stage rests on its distinctive faunal content rather than on the absence of trilobites.

The results of the analysis of the earliest Cambrian faunas and their possible antecedents can be summarized as follows. Five groups of taxa can be recognized among these faunas:

1. Groups first appearing in limited diversity but in large numbers, with recognizably primitive characters, subsequently going through successive adaptive radiations: Archaeocyatha, Foraminifera, Porifera. Their immediate ancestors were probably naked benthic protozoans.

2. Groups of metameric coelomates: ?Sabelliditida, Annelida, Arthropoda. Arising from the Late Precambrian coelomate radiation, represented in Ediacarian or older strata.

3. Lophophorates (oligomeric), represented by Brachiopoda which are believed to have originated from small, tubicolous burrowers. These are reasonably assumed to have been semi-sessile or sessile, with organic-walled tubes which are either unfossilizable or unrecognizable.

4. Mollusca, pseudometameric products of the acoelomate radiation. They are possibly represented in the Late Precambrian by trace fossils. The original adaptive radiation of the shell-bearing Conchifera was accompanied by a parallel but relatively short-lived radiation of sclerite-bearing, scaly and spiny Mitrosagophora and followed in the later Cambrian by another parallel development, the subphylum Polyplacophora (chitons).

5. The Hyolitha, Coleoloidea, and probably the Priapulida and the Chaetognatha have unknown or contentious phylogenetic relationships. The Hyolitha are considered as semi-sessile epifaunal deposit feeders, and this could apply to some extent also to other groups. Their ancestors would have been small, probably not fossilizable, worm-like animals.

This summary leaves only a small group of tubes, shells and sclerites of Early Cambrian age and small size in the '*incertae sedis*' group, particularly *Mobergella* and Bengtson's *Lenargyrion* (1977b).

The biohistorical problem posed by the events of the Precambrian–Cambrian transition is not so much one of sudden appearance of a diversified fauna as of the appearance of mineralized tissues in diverse groups of Metazoa in some abundance. Were there precursors?

### 4.5. Agglutinated, chitinous and mineralized body components in Ediacarian fossils and in some successors

The simplest method of constructing a hard covering of a soft animal body is by surrounding it with mud, sand grains, or mineral flakes from the environment. They are either loosely or firmly cemented with mucus secreted by glands in the integument. The single-celled Foraminifera include a number of taxa in which simple tubes or elaborate tests are constructed from various selected materials, including other foraminiferal tests, cemented with organic secretions from the cell content. Their fossil remains provide models for the fossilization of ancient Metazoa. Elongate conical or cylindrical tubes constructed from quartz grains have been found in the Lower Nama Group and described as *Archaeichnium haughtoni* Glaessner (1963, 1978). They were apparently rather loosely cemented, being twisted in places but coherent enough to be washed out of the sediment and redistributed on bedding planes by currents. I interpreted them as products of small, worm-like animals of unknown affinities. Cylindrical tubes of centimetre length and millimetre width, consisting of quartz grains and occurring in the lowest Cambrian, have been described as *Platysolenites*. Because they are almost indistinguishable from some species of the fossil and still living foraminiferal genus *Bathysiphon*, I consider them as the oldest Foraminifera (Føyn & Glaessner 1979). Like some agglutinated foraminiferal tests, the tubes of *Archaeichnium* were compressible and flexible, rather than brittle. Those of *Platysolenites* had thick walls with transverse growth layering. They tend to break across the tubes, parallel to these layers, but the tubes are mostly collapsed and flattened by compaction of the enclosing sediment. Flattening and distortion is often seen in younger fossil agglutinated Foraminifera in which organic cement is diagenetically hardened or replaced. Re-examination of the tubes of *Cloudina* from carbonate facies of the Lower Nama Groups (Germs 1972*a*, *b*, Glaessner 1976*a*) led to the conclusion that they were partly calcareous and secreted, with minor external agglutination of sand grains. Distortion of the tube walls suggests some flexibility 'reminiscent of some agglutinated Foraminifera with walls of quartz or carbonate grains and much organic cement'. Referring *Cloudina* to the Polychaeta, I noted that transitions from agglutination of detrital calcareous grains to cementation of secreted granules are known in the tube building of living polychaete annelids (Glaessner 1976*a*).

'Chitin biosynthesis... is a primitive characteristic of the animal cell – a property which has been kept during the evolution of the ancestral forms and most of the advanced ones in the protostomian lineage' (Florkin 1966, p. 53). According to Jeuniaux (in Florkin & Stotz 1971, pp. 595, 603) 'We can define a chitinous structure as a skeletal or cuticular structure which consists of chitin–protein complexes, that is to say glycoproteins or mucoproteins, the prosthetic group of which is entirely or principally chitin'. After defining it biochemically, he continues: 'The biosynthesis of chitin thus appears to be controlled by early established genes, present probably in the primitive unicellar root of Metazoa. This biosynthetic ability has been retained by a number of diblastic animals and by most of the triblastic Protostomia but was lost at the beginning of the deuterostomian evolutionary lineage.' It is hardly surprising that we find indications of the presence of chitin in the Ediacarian faunas. Chitin is highly resistant to decay, so that its former presence in fossil organisms can be determined by chemical tests. Regrettably, this has not yet been attempted in Precambrian fossils. Some of them have been described as being of 'chitinous' appearance, notably the toothed structures named *Redkinia* by Sokolov (1976*b*). M. A. Fedonkin kindly showed me a fossil consisting of a series of apparently 'chitinous', overlapping somites of an undescribed arthropod from collections near Arkhangelsk in northern Russia. The Pound Quartzite of the Flinders Ranges of South Australia contains no organic residues, presumably because it was deposited as a highly porous and permeable sand from which the organic matter was subsequently 'flushed' by circulating water. However, resistant fossil moulds such as the conularid *Conomedusites* and the chondrophoran floats represent structures which are chitinous in their later and Recent representatives. Similarly, the arthropods *Praecambridium* and *Parvancorina* show postmortem elastic deformations suggesting chitinous carapaces. The significance of the presence of chitinous substances in Late Precambrian Metazoa, still based only on indirect evidence, is twofold. It tends to confirm Florkin's conclusion which is derived from studies of the occurrence of this substance in living invertebrates in relation to their phylogeny. It is also related to mineralization of chitinophosphatic shells of inarticulate brachiopods and in some phosphatic tubular fossils or sclerites of Early Cambrian age which could have been comparable with Recent chitinophosphatic structures. The biochemical evolution of the cuticle of trilobites and other arthropods from Cambrian to Ordovician could give valuable quantitative indications confirming or refuting the impression that the integuments of some arthropods evolved from chitinous to dominantly phosphatic and later to more strongly calcareous mineralization.

Calcareous structural elements are probably not totally absent from the Ediacara fauna. Linear impressions within the composite moulds of the axis and polyp leaves of *Charniodiscus* and related colonial cnidarians could represent spicules which occur in the comparable tissues of living Pennatulacea. They are not sufficiently well preserved for detailed comparisons. Microscopic calcareous spicules are present in the tissues of many living Metazoa such as the Platyhelminthes, ascidians, or the subphylum Aculifera of the Mollusca. Lowenstam (1980) and Lowenstam & Margulis (1980) (see also p. 171) have indicated that they are entirely or almost unknown in the fossil record because of the difficulty of recognizing them among the calcareous debris in clastic carbonate sedimentary rocks, or because of their prevalent rapid disintegration.

The conclusion from this brief review is that the appearance of spicules and hardened, agglutinated, organic or biomineralized tubes or shells during the Precambrian–Cambrian transition was not an instantaneous or sudden event affecting different evolutionary lineages simultaneously, and so providing a simple biostratigraphic marker for a moment in geological time. It proceeded in rapid evolutionary processes from antecedents in Ediacarian time to its full manifestation in Cambrian and later times.

### 4.6 Extrinsic and internal factors of biomineralization and its functional significance

At the present state of our knowledge of Late Precambrian and Early Cambrian faunas, what requires explanation is not so much the sudden appearance of new kinds of Metazoa as the apparently rapid development of biomineralization processes in different phyletic lineages. Taking our guidelines from present knowledge of biomineralization processes in the living fauna, we examine a number of possible factors: (i) The influence of changes in the composition of the hydrosphere and atmosphere. (ii) The relevant physiological and biochemical basis of mineral metabolism. (iii) The biomechanical significance of 'skeletonization'. (iv) Selective advantages.

If the chemistry of fossil remains had changed suddenly (in terms of geological time) from total absence of pervasive presence of biomineralization, one would look for evidence of corresponding sudden changes either in the supply of essential bio-inorganic components, mainly phosphate, calcium, silica, or of necessary physical conditions, mainly suitable levels of acidity and perhaps temperature, in the marine environment. Because of the rapid exchanges between the atmosphere and the ocean water, one would consider also the possible influence of changed oxygen content of the atmosphere and, if possible, its carbon dioxide content at the relevant time. Not long ago, when data were even scarcer than they are now and speculation had few limits, it seemed that simple answers to these questions could be found. Since then, such concepts as the limeless ocean, or rapid changes in salinity or acidity in immediate pre-Cambrian time have been entirely or to a large extent eliminated from consideration. The original Berkner–Marshall theory placed the level of atmospheric oxygen at 1% PAL at the beginning of the Cambrian, which was thought to be the time of the first appearance of Metazoa. This view had to be modified. Cloud (1978b) now estimates that it amounted to about 8%. With this correction, it is still possible that increases in atmospheric oxygen level since that time have influenced evolutionary processes but it can no longer be maintained that a general threshold value was attained, with dramatic effect. The more subtle possible effects of atmospheric oxygen levels will be further discussed in the following section.

Fundamental physicochemical changes in the composition of sea water would have produced a recognizable difference between Precambrian and later marine sediments. There is little evidence of such differences producing a marked effect at the beginning of the Cambrian time. Geobiochemical and general evolutionary geochemical studies now in progress (e.g. Veizer in press) may lead to a reappraisal of more subtle effects. It should also be noted that Rutten (1966, p. 52) wrote of 'a change of pH of ocean waters leading to mild alkalinity and facilitating the construction of shells'. This was expressed in molecular terms by Degens, Johanesson & Meyer (1967, p. 640) who found that 'the Precambrian ocean was less alkaline than during later geologic times. A gradual increase in pH by a few tenths of a pH unit caused not only more favourable conditions for carbonate depositions but gradually opened the highly cross-linked proteinaceous matrix of the organic shell and made nucleation sites available.' Conditions for biomineralization may have changed but the essential substances used as building materials were available in abundance at all relevant times. Their geochemical abundance was such that tissue mineralization has been linked hypothetically with secretion, in the sense that 'skeletonization' may be thought to have an energy-saving selective advantage over excretion of a biologically overabundant component of sea water. This possibility was mentioned in connection with phosphatic shells. Their first appearance seems to have occurred during a period of unusually widespread and abundant phosphate mineralization. The Late Precambrian and Early to Middle Cambrian phosphate deposits are now being investigated by a Working Group of the International Geological Correlation Programme. It seems that more problems than solutions have so far emerged. The question of whether at times the organic component of the phosphate cycle dominated and contributed more residue to the sediments than at other times cannot yet be answered. In addition, there is the apparently unsolved problem of distinguishing biogenic from metasomatic phosphate in shells and skeletons by mineralogical or analytical methods.

At the beginning of the Cambrian Period there was little change in the silica cycle, except perhaps the slow beginning of the formation of siliceous tests in the radiolarian Protozoa. The greatest and most rapid increase was in the biomineralization by means of calcium carbonate. We must look beyond the surface of the 'shelly' animals to their physiology and biochemistry for ways in which those bio-inorganic substances are formed and utilized. Lowenstam & Margulis (1980), citing recent work by R. H. Kretsinger on molecular biology, state that intracellular regulation of calcium is crucial to the functioning of the eukaryotic cells. It is well known that calcium storage and release are essential to nerve and muscle activity; $Ca^{2+}$ and $Na^+$ extrusion evolved to 'protect' $Mg^{2+}$; calcium is linked to phosphate metabolism. Phosphates also act as inhibitors (crystal poisons) in calcite deposition which may require phosphatase enzymes for their removal. Kretsinger has distinguished calcium modulating proteins which maintain by means of their molecular conformation low concentrations of intracellular calcium and its 'informational' release for physiological processes in muscles and nerves. Other proteins, including extracellular ones, form matrices in or on which calcium mineralization, such as shell formation, occurs. According to Lowenstam & Margulis (1980), Kretsinger's first hypothesis explaining the important role of calcium in eukaryotic cells is that calcium carbonates and phosphates were first used as skeletal materials 'and then secondarily pumping and intracellular regulation evolved leading to its use as an informational system'. Since eukaryotes did not evolve from 'calcareous ancestors', this rightly seems 'highly unlikely' to Lowenstam & Margulis. They also refer to the Ediacarian animals which possessed muscle tissue and hence intracellular calcium regulation, a 'preadaptation that must have preceded rather than followed skeletonization'. Selection pressures towards its adaptive use are due to increased predation, the need for improved sensory and muscle systems for escape and, in sessile organisms, for rapid retraction into tubes. These physiological improvements were dependent upon calcium-modulating proteins and also provided calcium storage sites. The storage requirement is seen as one of the reasons for calcium secretion on the cuticles of early Mollusca. Lowenstam & Margulis do not believe that the fossil record reliably documents the onset of 'the innovation of biomineralization'. They refer to the fact that skeletal hard parts of marine organisms are often unrecorded from Recent and geologically young sediments in which they would be expected to occur. This happens either because they are dissolved in sea water or because they are taxonomically unrecognized and unassigned, or because they are disaggregated into single crystals of aragonite or calcite, being in living organisms mixed with extensive matrices of organic matter which are subject to decay. The hard parts may be destroyed by diagenesis, or by acid preparation of sedimentary rocks for the release of larger phosphatic shells. From these considerations, Lowenstam & Margulis conclude that the sudden appearance of certain Metazoa at the base of the Cambrian 'is more illusory than real', and that the suddenness 'is a consequence of dramatic differences in preservability of certain of the many stages prerequisite to the formation of calcareous skeletons'. This is a valid and valuable solution of one part of the problem.

The preconditions for biomineralization of tissues are inherent in the long processes of evolution of eukaryote cells and therefore of early Metazoa. Some, e.g. annelids, had acquired tubular shells in Late Precambrian time and some Cnidaria had spicules. Some biomineralized skeletal elements may still be unrecognized, for the stated taphonomic reasons. The other part of the problem, the reasons for the simultaneous commencement of diversification of shell and skeleton formation in Early Cambrian time in Archaeocyatha, Mollusca, Hyolitha, Brachiopoda and, perhaps later, in trilobites, requires an examination of the biomechanical significance of 'skeletonization' and of the selective advantages which stimulated adaptive radiations at that time.

A word of caution is required at this point of the enquiry. The problem of mineralized shells and skeletons, of their and their producers' diversification, is not the main or basic problem of metazoan evolution but only one of many. Calcium modulation and the management of other inorganic constituents of the environment are basic functions of the eukaryote cell physiology, including that of the Metazoa. The evolution of these functions does not necessarily lead to the production of mineralized, hard, protective shells or supporting skeletons. Soft-bodied animals live in large numbers and great diversity in the sea today, as they have done during the entire history of animal life. Many of them have given up ancestral shells or reduced their skeletons in the course of their evolution.

Zoologists use the concept of the hydrostatic skeleton. This represents a solution to problems of locomotion and other basic functions, an alternative to reliance on an interlocking system of hard parts (which is the original meaning of the word skeleton). What is required is not hardness but resistance to muscular tension. Such resistance can be provided by connective tissue built mainly from collagen, a structural protein. This contains a substantial amount of the amino acid hydroxyproline, whose synthesis requires molecular oxygen. Towe (1970, 1976, 1981) has expressed the opinion that a marked increase in atmospheric oxygen near the beginning of the Palaeozoic Era would allow the general utilization of this previously rare substance. This is described as the assigning of a higher priority to its use for a structural material, compared with its use for the purpose of respiration. This explains, according to Towe, the absence in the fossil record of the earliest, small and 'soft' lower Metazoa, animals with oxygen-limited collagen productivity, presumably mainly the Platyhelminthes and their predecessors. Therefore, the problem of skeletonization concerns the coelomate radiation and the appearance in the fossil record of the Ediacarian faunas. Whether it also explains the later (Cambrian) first occurrence of shells and skeletons – which presupposes the presence of abundant connective tissue in their producers – would appear to depend on quantitative data on rates of oxygen accession at that time, in relation to requirements for hydroxyproline production in soft-bodied compared with shell-bearing Metazoa. Such data are not available.

Vogel & Gutmann (1981) rightly point out that biochemical (physiological) and environmental (ecological) explanations are only contributions to the solution of the problem of the 'sudden' appearance of shells and skeletons but not the whole answer. Biomechanical and functional aspects are essential in the analysis of this evolutionary event. Skeletons, in the widest sense, are said to be formed in non-mobile parts of the body, otherwise its essential movements would be impeded. Annelid worms deform their total body mass in peristaltic locomotion. Their 'descendants', such as arthropods, exemplified by trilobites, can produce skeletons only when feeding and locomotive functions are transferred to ventrolateral limbs which reduce the need for total body movements. Energy is saved when the use of longitiduinal muscles for locomotion in conjunction with dorsoventral and circular body wall muscles is reduced by segmentation, together with the attachment of muscles to hard dorsal sclerites. This view of the annelid – arthropod functional transition differs from Clark's (1964) more detailed analysis mainly in the attention given to skeletonization, while his main concerns were the coelom and metamerization. In Vogel & Gutmann's view, similar biomechanical advantages, rather than those of protection, account for the origination of the molluscan shell. The transfer of the locomotive function to the foot makes movements of the dorsal area unnecessary. Calcareous deposits in the dorsal cuticle can then provide attachment areas for dorsoventral muscles, improving their efficiency. These primary sclerites are then fused to form the shell in Monoplacophora. An alternative view which conforms better with the stratigraphic distribution of early molluscan fossils is that of Runnegar (1980b) and Pojeta (1980) who invoke partial parallelism between mono- and polyplacophoran shell formation rather than fusion of valves of an hypothetical polyplacophoran ancestor. In Brachiopoda (Gutmann et al. 1978), shells originated in the collar area of phoronid-like, burrow-dwelling ancestors which were postulated also by Wright (1979). This could occur when longitudinal total body movements were no longer required. When fully developed, these shells improve water flow to the lophophores and protect them. Vogel & Gutmann (1981) apply similar considerations not only to arthropods and molluscs but also to the skeleton formation in the Chordata (but this is probably not relevant to the events at the Precambrian–Cambrian transition). The authors state that the evolutionary processes described also lead necessarily to the development of a respiratory system replacing diffusive oxygen intake through the entire body surface. The emphasis placed on changes in locomotion in the early stages of arthropod evolution is used to explain the supposed absence of trilobite dorsal exoskeletons in Tommotian sediments. It could also explain the presence of trilobite-like locomotion traces (*Rusophycus*) in strata of that age. It could be assumed that sclerotization of arthropod legs preceded that of the dorsal exoskeleton. In Vogel & Gutmann's opinion the ancestors of the Mollusca and Brachiopoda were not only soft-bodied but small and therefore they are not likely to be found as fossils. The appearance of the Cambrian 'shelly' fauna is considered to have been even more gradual than the fossil record suggests. The structural plans ('Baupläne') of the Cambrian Metazoa are not new but modified by skeletonization for which they must have been preadapted morphologically (and also physiologically and biochemically, as Towe and Lowenstam & Margulis have shown).

Most observers of metazoan organization and evolution agree that the potential for biomineralization is an essential basic property of the eukaryote cell. At the stage which eukaryote evolution reached during the Precambrian–Cambrian transition, when shells become manifest in the record, their functions go far beyond that of protection from predators. The naive assumption that shells are acquired because they protect soft bodies seems influenced by anthropocentric thinking: man uses shields for protection from aggressors. Sophisticated analyses have shown that protective functions of shells, sclerites and skeletons are often secondary to others such as: energy conservation, as fixed muscle attachments replace anatagonistic muscle action; support, including raising of the body above the sea floor for gathering and filtration of organic matter before it settles; and built-in direction of water currents containing nutrients and oxygen, to the animal's advantage. Recent work on Late Precambrian faunas (see preceding sections) and taphonomic considerations have reduced the apparent suddenness of the biomineralization event at the beginning of Cambrian time. Some of it has been explained as a grade reached as a result of selection pressures for increased bioenergetic efficiency in early Metazoa. When and where it occurred, it was controlled by biomechanical constraints. A residue of problems remains which requires examination of possible external, environmental and palaeoecological factors.

## 4.7 Environmental changes at the Precambrian–Cambrian transition

In recent years the rapid increase of palaeontological data for the period here considered led to the realization of the complexity of factors that brought about a peculiar succession of peculiar faunas. It became clear that no single change in one of the physicochemical parameters that control animal life in the sea could be held basically responsible for what had at first appeared to be a critical phase in the history of life. With more biostratigraphic evidence, events which seemed to mark a critical crossing of a boundary were found to have occurred within millions, perhaps tens of millions of years. The outstanding event in Late Precambrian time was a series of major glaciations but that had occurred 10, 20, if not 100 million years before the main transformation and the dispersal of the new fauna. It may have had after-effects which will be examined later, but it could not have been the main factor introducing the Phanerozoic era in the history of life.

Hypothetical changes in salinity or composition of sea water have not been observed in the geochemistry of relevant sedimentary rocks. Biological changes in the organisms were found to be more plausible explanations than changes in their inorganic environment. The search for a single extrinsic factor explaining biohistorical events at the Precambrian–Cambrian transition is no longer appropriate. The assumption is unavoidable that life was influenced then as it is now by global changes in its environment. Significant changes which could have steered the course of evolution at the time here considered have been briefly noted in the preceding account. One of them is the accelerated change in the environment of marine life which must have resulted from the sequence of major transgressions and regressions that characterizes the beginning of Cambrian sedimentation in many regions. The other is the generally accepted increase in atmospheric oxygen from Early Precambrian time. Oxygenic respiration must have been possible long before the beginning of the Cambrian. If the oxygen content had by then risen not to $1\%$ but $8\%$ PAL, its further increase, perhaps to $10\%$ by Siluro-Devonian time, about 400 m.y. ago, could have affected metazoan evolution as much as the preceding rise. A third global environmental effect which has not been disproved and therefore should be considered is a possible rise in temperature. It should not be left out of consideration because the main remaining problem is the rapid appearance of mineralized tissues in diverse unrelated groups of fossil Metazoa which are not found in earlier sediments. It is true, as stated by Stanley (1967a, p. 73) that 'Major skeletal taxa appeared sequentially, not simultaneously' but is it also true that 'The appearance of a skeletal record in the Cambrian was simply part of the general diversification and requires no special explanation'? When environmental changes coinciding with the Precambrian–Cambrian transition are examined, it may still be worthwhile asking which of them, if any, could have brought about a sudden increase in diverse forms of biomineralization.

Palaeogeographic changes, documented for the relevant transition period by transgressions, should be considered because they must have offered new habitats with advantages for skeletonized, shell-bearing animals. Stepwise increase of oxygen supply (undocumented for the relevant time interval) may be significant, in accordance with Towe's reasoning: it may be significant for an evolutionary step, in this instance biomineralization, which requires more oxygen for its prerequisite, i.e. collagen production. Temperature increase, also otherwise undocumented, may facilitate calcification. These physical factors are discussed at this point not in order to hold one or the other responsible for what has happened. Every evolutionary event depends not only on the physical but also on the biological environment. Both kinds of environmental factors would have influenced the events which we are considering. This interaction is well shown in a model (Fig. 4.5) constructed by Brasier (1979). He states (p. 146) 'The Cambrian radiation of invertebrates and the transgressions are...almost inseparable'. The major transgressions are generally held to have made new, highly productive, shelf areas available to the marine fauna. This was particularly significant in Early Cambrian time when many ecological niches were still effectively unoccupied and, as Sepkoski has shown, an exponential taxonomic diversification was in progress. Brasier (1979, p. 144) explains his model as follows:

> The succession of dominant environments through the transgression may have controlled the timing of the development and radiation of different fossil groups. In this model, littoral communities would develop during the initial stage of rising sea level. Rapid transgression allows them to spread more widely (the 'radiation stage') whilst a contemporaneous assemblage develops further offshore. If the transgression proceeded in continuous pulses, the story of the Cambrian radiation would have approached the model..., with the evolution of different communities out of phase with each other. New stocks evolved rapidly during their development stage because they were small populations in isolated habitats. These habitats were destined to expand with time allowing global colonization, diversification and evolutionary radiation.

We need not accept all details of Brasier's model and its explanation to gain insight into effects of environmental changes during Early Cambrian transgressions. The model can be easily modified to fit specific tectonic and geographic conditions in different regions. It is important to keep in mind Brasier's conclusion that the Cambrian adaptive radiation is 'overprinted by an element of ecological succession' because to some extent biofacies must shift with lithofacies during times of transgressive–regressive events. In modified forms, similar models would fit any other major transgressive–regressive phases with global effects in geological history. The more fundamental and more obvious phylogenetic effects of the Cambrian

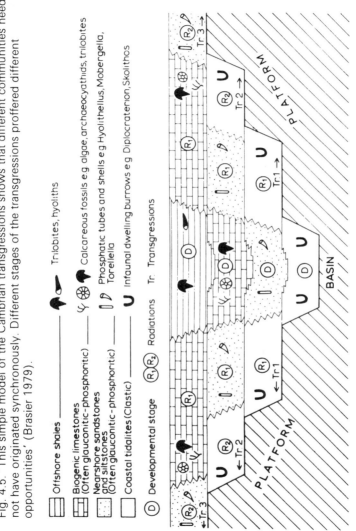

Fig. 4.5. 'This simple model of the Cambrian transgressions shows that different communities need not have originated synchronously. Different stages of the transgressions proffered different opportunities' (Brasier 1979).

transgressions are distinct from similar events at other boundaries of eras owing to the fact that this historical phase is not – as the later ones are – a time of replacement of pre-existing metazoan communities. Such communities existed, as shown by the Ediacarian faunas and by their 'prehistoric' antecedents. They were limited by a number of factors which deserve consideration even though they are hypothetical rather than strictly demonstrable on presently available data. Firstly, to explain the point by an analogy from economic history, at the time of the Cambrian transgressions, animal communities were 'developing' rather than 'developed', as we say today referring to various human socio-economic communities. Secondly, the total animal biomass appears to have been significantly smaller than at later times. Perhaps it is no accident of preservation or discovery that the oldest exploitable deposits of hydrocarbons are of Vendian–Early Cambrian age and few in number, just like the oldest known major deposits of animal fossils ('Fossil-Lagerstätten'). Thirdly, major environments and components of later biota had not yet evolved. Examples are reefs; stromatolite bioherms and biostromes existed but had yet to be 'cropped', with consequences that were elucidated by Awramik (1971) and Garrett (1970) in relation to stromatolites and by Stanley (1976b) with reference to metazoan diversity. Preferential occupation of certain habitats by primitive Metazoa was deduced from observations on the living fauna. Oxygen-deficient environments were studied from this viewpoint by Rhoads & Morse (1971) and by Fenchel & Riedl (1970). Burrowing habits were considered responsible for the innovative evolution of a coelom and metamerism (Clark 1964, 1979). The implication that other habits and habitats are therefore excluded from influencing the dispersal and diversification of early Metazoa is not in accordance with the geological and palaeontological data. Nonetheless, the conclusion can be supported that during the Precambrian–Cambrian transition an expansion of metazoan living space took place. Many new ecological niches were occupied by the evolving Metazoa, without replacement of the pre-existing biota. It is not possible to specify the new niches resulting from the Early Cambrian transgressions when discussing events on a global scale. Brasier (1980, 1982) has referred to specific examples in the Lower Cambrian of England and Norway and there are others from Scandinavia and the USSR. Their discussion would exceed the limits of this chapter. Even a complete listing of all known Precambrian–Cambrian stratigraphic sequences and their faunas would not disclose all kinds of animal communities or ecological niches that were established during this transition period. It was shown in Chapter 2 that the few known collecting localities of Ediacarian faunas represent very different lithofacies, from quartzites to siltstones, tuffaceous argillites and limestones. All are neritic, more or less shallow water facies. Whether or not the continental slopes (bathyal) or the deep ocean floor regions were inhabited in Vendian time is unknown because we do not know unmetamorphosed oceanic sediments of that age. It is likely that they would not be recognized if found, because the distinctive micro- and nannoplankton had not evolved until later times, or had not evolved mineralized (calcareous or siliceous) or resistant and distinctive organic integuments. Moreover, most if not all deep-water sediments formed off continental margins would have been metamorphosed in subsequent orogenies or subducted below the continental crust. Intracratonic basins contain in some rapidly sinking zones (aulacogenes of the East European Platform and other regions) well-preserved Precambrian and younger sediments but there seems to be no evidence that any of them were deposited at bathyal depths.

The Middle Cambrian Burgess Shale (British Columbia) with its mainly soft-bodied fauna was deposited in relatively deep water. It is considered to have lived in depths of over 100 m (Whittington, 1980) and to have been deposited at 160 m (according to Fritz, cited by Conway Morris 1979*b*). It is unknown what part, if any, of the Late Precambrian–Cambrian faunas lived at greater depths than the edge of the continental shelf. Its configuration at the time here considered is unknown, as is the general physical and biological palaeo-oceanography. As stated before, the Early Cambrian transgressions must have increased the shelf areas and the available habitats and niches on them. They must have been then, as they are now, the most favourable areas for metazoan diversification because primary production of nutrients by plants depends on light penetration and the depth of the photic zone is comparable with that of the shelf. Wave action of the shelf must have contributed significantly to mixing of atmospheric oxygen and sea water.

Weyl (1968–69) has indicated the significance of the thermocline at low latitudes as a possible environment for the evolution of eukaryotes. If the photosynthetic oxygen-producing blue-green algae were restricted to the thermocline by density constraints, early animals would have been originally planktonic consumers (zooplankton). The later development of the ozone layer would have allowed formation of an algal cover on the sediments of the shelves, in a shallow seasonal thermocline with temporary and still local concentrations of oxygen until its atmospheric concentration became 'sufficiently high'. At that time, the beginning of the Cambrian according to Weyl, the animals could become benthic and increase their bulk density by skeletonization. Hobson objected to this hypothesis, citing data against concentration or confinement of particulate matter by the thermocline, but Weyl cited other data and there the matter appears to have rested. The question of 'layered' oceans remains open until the palaeogeography and palaeo-oceanography of the Precambrian–Cambrian transition becomes clearer; it can be of considerable importance for the early evolution and distribution of animals in the sea. What is significant for the question here discussed is the possibility of a mechanism in the ocean linking 'sufficient' oxygen production with a transition from planktonic to benthic animal life on the continental shelf and relating development of shells to this transition. In view of the existence of benthic Ediacarian animals, the time of 'sufficient' oxygen in the atmosphere must have been Precambrian. An increase in ocean temperature at the beginning of the Cambrian to allow shell formation remains speculation. Those who do not mind speculating may think of the possibility of linking a rise in global temperature with increased carbon dioxide content in the atmosphere, producing the often invoked greenhouse effect. In turn, it could be linked with greater volcanic activity at a time of widespread orogenic events (Pan-African orogeny, Baikalian orogeny in Asia, Delamerian orogeny in South Australia and Antarctica). To come back to earth, or rather to the ocean: what is required for fuller understanding of environmental changes influencing evolutionary events at the Precambrian–Cambrian transition is a knowledge of physical, chemical and biological oceanography of that time. This is still lacking. Essential to the consideration of environmental changes is the biological environment, the synecology, or in modern terms the changes in community structure. Unfortunately, factual data have been examined for the critical period in only a few pioneer studies (see particularly Brasier 1982). There is no lack of theories explaining what could have happened but the time has not yet come to summarize and analyse sufficiently fully data on the changes in community structure at the relevant times. The frequently made assumption that large-scale predation evolved then and led to the rapid production of protective shells touches on only one of the many facets of the problem, from a singularly anthropocentric viewpoint – which is perhaps understandable at a time when the arms race and defence expenditure are daily news items.

### 4.8 Increase in diversity of trace fossils

In the context of the study of the Precambrian–Cambrian phase in the evolution of the Metazoa, particularly their initial diversification and dispersion, the contribution made by trace fossils requires evaluation as a special kind of document of the presence and activities of animal life. In this context it does not require description or comprehensive listing of these fossils. Following the lead of Cloud (1968, 1973) and the efforts of Hofmann (1971, 1972), objects of non-biogenic, doubtful, or Recent origin will not be discussed. It is necessary to determine what the animals were doing to and in the sediments and what organs they could have used for their recorded activities. This would give some information about the level of their evolution and may indicate what kinds of animals could have done what is recorded in the rocks. We would then know what morphological equipment was available and what ethological conditions were required for making the traces. Due consideration would have to be given to their fossilization and particularly to later diagenetic changes. Regrettably, little of this information is at present available. A thorough sedimentological study of the trace fossils as well as their matrix is needed. In the present context it is not necessary to evaluate the stratigraphic or the detailed palaeoecological significance of specific Precambrian–Cambrian trace fossils which, with a few exceptions, still require more work in a regional context before they can be used to support general conclusions.

Named taxa of trace fossils are based on the morphology and topology of traces and at best on only a minor portion of the morphology of the animals (size, symmetry, legs and other means of locomotion). They are fundamentally different from taxa in the zoological system and therefore, in studies of diversity, their number should not be included in quantitative reviews of Precambrian–Cambrian biotas and counted with other taxa. This can only be done meaningfully if the number of trace fossils rigorously excludes those made by the same animals in different circumstances or by those already counted as body fossils

A number of trace fossils occurs with the Ediacara fauna in the same strata (see p. 65). Six were distinguished as Forms A to F and five of them were figured (Glaessner 1969a). All were made by worm-like, or more precisely, worm-shaped animals. Many living marine animals are worm-shaped but differ widely anatomically and are therefore assigned to many different phyla. It is impossible to assign the Ediacarian trace fossils to definite places in the zoological system. To group them all together as 'Vermeidea' (Zhuravleva 1970), together with worm-shaped tube dwellers, has no explanatory value. There were benthic animals that burrowed through sediment, but dwelling burrows such as *Skolithos* or *Diplocraterion* are absent. In the Pound Quartzites *Diplocraterion* burrows occur only below their disconformable or unconformable contact with the Lower Cambrian Parachilna Formation. They were made at the time of the deposition of that formation, prior to the diagenetic hardening of the Precambrian sand.

Germs has now reported *Diplocraterion* from the Lower Nama. In view of the abundant evidence for its Ediacarian age he concludes that this trace fossil was not absent from but rare in the Ediacarian trace fossil assemblages, but its illustration (Crimes & Germs 1982, which appeared too late for detailed discussion) is unconvincing, as are other classifications in this paper. Only examples of trace fossil assemblages from selected regions will be discussed and their use as a basis for stratigraphic correlation will be disregarded. The Late Precambrian trace fossils from western New South Wales will be mentioned because their correlation with the Ediacara assemblage has caused some misunderstandings. Webby (1970) described two trace fossil assemblages from two formations in the Upper Torrowangee Group of western New South Wales. The older one comprises a few branching or winding 'worm trails', 1–3 mm wide. The younger assemblage from the Lintiss Vale beds, about 1500 m higher and at the top of the group, includes *Planolites*, *Cochlichnus*, *Torrowangea*, *Phycodes*? *antecedens*, and *Curvolithus*? *davidi*, a locomotion trail consisting of three longitudinal bands, 2–6.5 mm wide and separated by narrow grooves. The older assemblage may have contained three to seven different patterns of activity by small, mainly infaunal, sediment feeders. The younger assemblage consists of 10–21 patterns of dominantly feeding burrows, a few rest marks and crawling trails. Apart from millimetre-deep vertical *Planolites* burrows, virtually all trace fossils follow bedding planes. Webby considered both occurrences to be of Precambrian age. However, Daily (1973) correlated the upper, richer assemblage with the Uratanna Formation of Early Cambrian age. This correlation was accepted by Webby (1973), although in his view the only common ichnogenus was *Curvolithus*; he preferred to place the Precambrian–Cambrian boundary above either one or preferably two of Daily's Lower Cambrian formations (hence approximately at the top of the Tommotian). Accordingly, the 'abundant and diverse' trace fossil assemblages of central Australia are in his opinion of Vendian (Ediacarian) age, a view which is not in agreement with observations made elsewhere. Although the assemblages from western New South Wales lack such distinctive Cambrian (Alpert 1977) ichnogenera as *Diplocraterion*, *Plagiogmus* or *Dimorphichnus* and the upper one is not as diverse as the middle and upper 'Arumbera' assemblages from central Australia, they do not resemble the Ediacara ichnofauna particularly closely. They could conceivably be slightly younger, corresponding in age to the pre-Uratanna unconformity of South Australia or to the latest fossiliferous Precambrian of central Australia. Further ichnological discoveries could, with careful analysis, test this hypothesis of a distinctive latest Precambrian ichnofauna.

The trace fossils of the richly fossiliferous Vendian deposits occurring on the northern and western margins of the East European Platform were described by Fedonkin (1976, 1981*a*) and by Palij *et al.* (1979). There are feeding traces (*Neonereites, Harlaniella*), locomotion trails (*Bilinichnus* and *Cochlichnus*) and a supposedly bilaterally symmetrical resting trail (*Vendichnus*) which is 10–35 mm long and 8–20 mm wide, with what is described as traces of four to five pairs of appendages. There are also the 'grazing trails' (*Palaeopascichnus*), consisting of transverse, short ridges and grooves, and the similar but sinusoidally arranged *Nenoxites*. Both could be fecal trails, similar to *Neonereites*. To this category the authors added *Suzmites volutatus* but it is now considered as '*incertae sedis*'. It could have been left by a *Pteridinium*, a metazoan taxon, remains of which were found nearby in a vertical position (Fedonkin 1981*a*). Its movement over a bedding plane could have been induced mechanically. The uncharacteristic 'worm trails', commonly named *Planolites*, also occur in this assemblage, closely resembling form E from Ediacara. A 'resting trail of a small bilateral animal' (Fedonkin 1977) appears to be an impression of a *Charnia* or *Charniodiscus*. As at the Ediacara localities, no dwelling burrows have been observed. This is significant as the sediments were laid down in a lower energy environment than the Pound Sandstone and therefore the evolutionary level of Ediacarian faunas rather than facies may be responsible for their absence. Fedonkin (1977, 1978*a*) noted that in the Precambrian fauna the behaviour pattern of the producers of trace fossils was such that essentially only two-dimensional traces were left 'as the animals moved only near the sediment/water interface'. The intricate three-dimensional burrows of Palaeozoic times are absent, as are vertical dwelling burrows, but the animals were able to move down through nutrient-poorer sand to more nutritious buried clay layers, producing endogenic hypichnial traces.

On the northern margin of the Baltic Shield, in Finnmark, Banks (1970, 1973) studied the distribution of trace fossils in a continuous section from the Upper Precambrian to the Lower Cambrian. He noted an increasing abundance and diversity which he ascribed principally to the evolutionary development of the Annelida, Arthropoda and Mollusca. Sedimentary facies cannot be held responsible for the overall increase in abundance but the increase in atmospheric oxygen was considered to be a fundamental factor. Stratigraphic correlation was extended along the front of the Caledonides to southern Norway (Føyn & Glaessner 1979) and the uncertainty of Precambrian or Early Cambrian age was reduced to the upper half of the Stappogiede Formation in the north and the Vardal Sandstone in the Mjøsa District in the south.

In North America, much work on trace fossils from the Precambrian–

Cambrian transition remains to be done or published. The main conclusions from an unpublished thesis were given by Alpert (1977). Twenty-six genera of trace fossils including six different 'trilobite or arthropod traces' were distinguished as indicative of Early Cambrian age. Seven named and several unnamed dwelling (*Skolithos*), feeding (*Planolites*) and locomotion trails, the latter including '*Scolicia*', *Curvolithus*, *Didymaulichnus*, *Torrowangea*, *Helminthoidichnites*, etc., were recorded from Late Precambrian and basal Cambrian strata. References to Precambrian *Skolithos* have since been considered as erroneous but the Precambrian and Cambrian occurrences of the other genera have been confirmed. It is the opinion of several observers that trace fossils reflect the evolutionary development of Metazoa and their behaviour during the transition time. Their increase in abundance, diversity and complexity of behaviour has been clearly demonstrated worldwide. The simplicity of the Late Precambrian trails compared with the more elaborate Cambrian and younger burrows and feeding traces has been noted. The relation between these unquestioned observations and the supposed dramatic increase in rates of evolution for which they were considered as 'compelling evidence' (Osgood in Frey 1975, pp. 104–5) is another matter. This view may be based on rates of taxonomic diversification of the Metazoa but the evidence from the increasing diversity of trace fossils is weakened by the difficulty of assigning different successive kinds of trace fossils to emerging classes or phyla. Earlier trace fossils as well as many later ichnogenera represent worm-like animals. Traces of annelid worms can be distinguished only exceptionally from those of very different kinds of 'worms'. Amphipod crustaceans can make similar burrows and Mollusca can make superficially similar locomotion trails to those of worms in Recent shallow water environments. Fecal castings of sediment-feeding Hemichordata (Enteropneusta) cannot be distinguished from those of holothurians on some sand flats (Howard in Frey 1975, p. 142, Fig. 8.6). The mechanics of locomotion applied by different major taxa of Metazoa have been considered in the assignment of trace fossils. The lever movements of arthropod legs leave different imprints from those of legless worms which move by peristalsis. Mainly because of the small size of most parapodial setae, characters distinguishing traces of annelid parapodia from those of small arthropod legs are not understood. *Diplichnites*, *Rusophycus* and *Cruziana* are certainly traces of arthropods and some of them are known to have been made by trilobites. Gastropods and their molluscan ancestors are characterized by a muscular foot. They may also have left traces of their mantle fold on the ground over or in which they moved on a bed of mucus. It has been suggested that complex trails found in Upper Precambrian and Lower Cambrian strata could be ascribed to Mollusca. One of them is *Didymaulichnus* (Glaessner 1969a, Young 1972), a bilobed, medially grooved trail with, in some places, bevelled lateral bands (see p. 129). Another trail (endichnial) which has been described as 'molluscan-like' is *Buchholzbrunnichnus* from the Lower Nama Group (Kuibis Subgroup) of southwest Africa (Germs 1973b). It has transverse, close-set ridges and grooves on one face and a smooth, trilobed profile on the other. As it was found on a loose slab, its orientation is unknown but it resembles the Lower Cambrian *Plagiogmus* (Glaessner 1969a, Kowalski 1978, Jaeger & Martinsson 1980) which has widely spaced transverse ridges ventrally and a subcentral longitudinal furrow in its smooth dorsal face. There is also a remote resemblance to the giant Late Cambrian trail *Climactichnites*. The systematic position of the originator(s) of these trails is unknown. Presently unknown metazoans, with unexplained mechanisms of locomotion, have left traces in and on Late Precambrian–Cambrian clastic sediments.

From the higher part of the Lower Nama Group, mainly the Nasep Quartzite Member, Germs (1972c) described a number of trace fossils, most of which also occur elsewhere. They include *Planolites*, *Torrowangea*, *Didymaulichnus* (described under other names), and three-ridged trails. *Archaeichnium haughtoni* Glaessner, wrongly reported from the Kuibis Quartzite, was actually found in the Nasep Quartzite. It is not a trace fossil *sensu stricto* but represents worm tubes, partly agglutinated and partly organic, which had been washed out of the sediment and distributed on bedding planes. At the top of the Lower Nama, above an unconformity, and in the succeeding Fish River Subgroup, above another unconformity, Germs found *Phycodes pedum*, which has a wide distribution in the Lower Cambrian.

The lowest Cambrian (Tommotian) from the margins of the East European Platform contains, in addition to genera already mentioned, the trace fossils *Treptichnus* and *Teichichnus*, which have composite and easily recognizable burrows, and *Bergaueria*, commonly believed to be a dwelling burrow of a sea anemone, appears in the Lower Cambrian in Europe and North America.

The functional–morphological or biomechanical analysis of the emerging trace fossil complexity is much less advanced than its ethological, ecological and biostratigraphic investigation. An example of a biomechanical approach is seen in studies of the changes in *Cruziana*-like arthropod locomotion trails which were partly or mostly made by trilobites. Crimes (in Frey 1975) sees evidence for a progression from simple furrows (earliest Cambrian) made by shallow, low-angle, slow movement produced by a simple muscle system with low efficiency. Rapid evolution produced a change from a soft to a hard, sclerotized or mineralized integument on which muscles could work more efficiently. These arguments were later elaborated by Vogel & Gutmann (1981). Crimes links this evolutionary advance with atmospheric oxygen increasing above a 'threshold value'. This follows the reasoning of Rhoads & Morse (1971, Rhoads in Frey 1975). In present marine environments where oxygen falls below 1 ml/litre, the sea floor is dominated by small infaunal deposit feeders (see also Raff & Raff 1970). As oxygen falls below 0.1 ml/litre, most metazoans disappear. According to Rhoads & Morse, this corresponds to 1–2% of the present amount of dissolved oxygen or 0.01 of the present atmospheric level (1% PAL or the Pasteur Point). They believed that this level was reached 700–800 m.y. ago. Taking modern dysaerobic basins (0.3–1 ml/litre) as a model for the Precambrian–Cambrian transition, they calculated atmospheric oxygen at about 10% PAL at the time of the first appearance of diverse calcareous faunas. A lesser value of 6–7% PAL was subsequently suggested by Cloud (1976a) for the time of the first appearance of Metazoa at about 700 m.y. The likelihood of a threshold value having been reached at that relatively high level (high in relation to low oxygen levels tolerated by certain living Metazoa) was correspondingly reduced. Besides, neither trace nor body fossils of the Ediacara fauna are small (up to 15 mm width of trails, over 500 mm length of body fossils), although there are wider trails in the Lower Cambrian. On present knowledge, however, increased oxygen availability cannot be ruled out as a factor contributing to greater diversity of Cambrian trace fossil makers. Another factor is selection for greater economy of effort in feeding and locomotion. As most of the trace fossils of the Precambrian–Cambrian transition sediments are feeding and locomotion trails, a wide field of investigation of changing mechanisms is still open. Regrettably, little has been done to distinguish between open tunnels, mucus-cemented burrows, traces of sediment-ingesting fauna, selective detritus feeders, and filter feeders, to name only a few types of concomitant locomotion and feeding traces that could have been preserved. They await investigation by sedimentological methods, particularly by thin-section studies. Their aim should be the careful description and interpretation of the matrix and infilling of burrows. Active biogenic or subsequent mechanical infilling should be distinguished, effects of differential diagenesis recognized, and wall cement and fecal matter analysed. Such studies are particularly needed for Precambrian–Cambrian material because it is probable that apparently similar Ordovician to Recent traces, some of which have been studied in this manner, were actually produced by different animals in a different manner. Classification under identical names indicates only topological and through it possibly ethological or ecological similarity but rarely the biological identity of the animal. This identity can be deduced from indications of its muscular efforts in relation to the sediment, mucus production, food acquisition and waste excretion. All or some of these points could be elucidated by the study of sedimentary particles relocated, reworked or ingested by trace fossil producers. The palaeophysiological and autecological results of such investigations would add greatly to the studies of the evolution of trace fossil communities (see Seilacher 1977, and his numerous previous publications).

Most of the Late Precambrian and Early Cambrian trace fossils were made by infaunal and epifaunal detritus feeders, either ingesting sediment or selecting organic particles from it. In the Lower Cambrian sediments they become more abundant and more diversified. The earlier forms are joined by vertical straight (*Skolithos*) or U-shaped (*Diplocraterion*) dwelling burrows or others probably made by sea anemones or similar cnidarians (e.g. *Bergaueria*). Arthropod traces appear for the first time, either walking tracks such as *Diplichnites* or *Cruziana* or 'resting traces' which might be more appropriately named digging traces (*Rusophycus*). In many Palaeozoic stratal sequences, poorer and simpler trace fossil assemblages are followed by richer and more diverse ones. An example of this was given by Crimes *et al.* (1977) who concluded (p. 129): 'The marked facies control shown by many trace fossils will make it difficult, but not necessarily impossible, to use them to define a base to the Cambrian System and to accurately evaluate the hypothesis of explosive evolution at that base'. We have discussed the difference between defining a base to the Cambrian System and evaluating the events during the Precambrian–Cambrian transition (p. 139). This evaluation requires a clear distinction between the term explosive evolution and the dramatic increase in rates of evolution for which trace fossils of that age are thought to give compelling evidence. We know too little about the identity of trace fossil producers to find numerical evidence for an increased rate of diversification beyond that already available from body fossils. The first appearance in the Early Cambrian of new trace fossils such as those listed above is a qualitative if not quantitative confirmation of what is generally understood as 'explosive evolution'. This means no more than an indirect confirmation of, or rather conformity with, the exponential rate of evolution from Vendian through Early Cambrian (Sepkoski 1978, 1979, Brasier 1982). When Clark (1979) says that in the early coelomate radiation the unsegmented, metameric and trimeric types of coelomic organization appeared as different adaptations to life in, rather than on, the substratum, this is in agreement with what is known about Ediacarian trace fossils. Their producers lived mostly in the upper layers of sediment, seeking nutrients from detritus, or burrowing through these layers to search for recently buried, nutrient-containing layers at relatively insignificant depths. Endobenthic annelids were among the first fully mobile metazoans (Frey & Seilacher 1980). It is probable and widely accepted that molluscan ancestors (pseudometamerous), leading a similar life at that time, are known only as trace fossils. In Early Cambrian time some burrowers (in the widest sense of the term as adopted by Clark 1964 and 1979) evolved their biomechanical capabilities to emerge in various new ways: as sessile dwellers in shallow holes (*Bergaueria*) or tubes (*Skolithos*) from which they could gather food on or above the sediment surface or in which they could contract or retract; or as mobile arthropods, capable of active search for food; or as advanced filter feeders, raising themselves in or from tubes above the sediment surface to extract nutrition from the water. That much can be confirmed, to a significant extent, from the study of trace fossils.

Fedonkin (1977, 1981*a*) considers that many trace fossils representing locomotion and feeding activities consist topologically of elementary patterns such as repeating or alternating right and left turning, sinusoidal or spiral, branching or radial, basic units. Trace fossils found in Vendian sediments are close to the elementary patterns and are essentially two-dimensional, as mainly the surface muds were being exploited. The Lower Cambrian assemblage becomes more complex topologically. There are branching patterns and three-dimensionally stacked and spiral burrows. Their complexity grows in later Phanerozoic times. It is due not only to selective optimization of exploitation of resources but also to changes in condition and rates of sedimentation. Improvements of locomotive, ingestive and sensory organ systems should also be considered. They may be preadaptive for the construction of burrows and other potential trace fossils.

### 4.9 Consequences and causes of evolutionary diversification

Metazoan diversification from Late Vendian (Ediacarian) to Middle Cambrian time created a qualitatively and quantitatively distinctive assemblage of taxa. This historical sequence is known, according to different viewpoints, as the Cambrian radiation event or the origination of the 'Cambrian fauna' (Sepkoski 1979). Sepkoski's quantitative analysis of emergent faunal diversity (Fig. 4.6) has the merit of documenting and attempting to explain the distinction between the 'Cambrian fauna' and the subsequently emerging 'Palaeozoic fauna'. The word 'fauna' means in this context not the total diversity of animals living at one time in one place or region but the total emerging diversity of animals which is either maintained or declines through the interaction of rates of origination and extinction. The diversity data are presented by Sepkoski at the level of families. His models and their discussion show clearly that the exponential or so-called explosive diversification during Early Cambrian time did not directly lead to the steady state or equilibrium of the Palaeozoic fauna. The analysis reveals a two-stage process in which a 'pseudo-equilibrium' was reached in mid- to Late Cambrian time. The origination rate increased again in the later part of the Cambrian, leading in the Ordovician to a threefold increase of familial diversity compared with that at the Cambrian apparent equilibrium. This diversity is maintained to the end of the Palaeozoic, with fluctuations marking period boundaries. This statistical evaluation accords well with the empirical conclusion reached by observers who are familiar with Mesozoic and Cainozoic faunas and who find Ordovician associations of fossils relatively familiar. They are largely

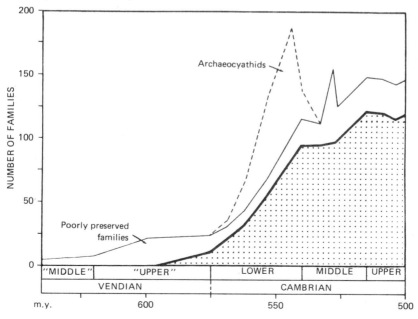

Fig. 4.6. 'The known diversity of metazoan families within the Vendian and Cambrian plotted against geologic time, illustrating the basically logistic pattern of diversification during the interval from 650 to 500 m.y. The stippled field represents the diversity of 'normal', heavily skeletonized families, exclusive of archaeocyathids, whereas the open field above represents the additional diversity contributed by families based on soft-bodied or lightly skeletonized fossils. Numbers of families of archaeocyathids are depicted by the upper dashed line; this group does not fit the logistic pattern... The peak in poorly preserved families over the mid-Middle Cambrian is caused by the Burgess Shale fauna' (Sepkoski 1979).

interpretable in terms of the living fauna, both as to functional morphology and ecology. Turning to faunas from the Lower to Middle Cambrian, most of them will be found too strange to permit ready placing of taxa in the system of living animals. Many of them will be classified as of unknown affinities. Likewise, there will be much greater difficulties in establishing their habits and habitat characteristics. One will look in vain for the familiar coral reef associations, finding only remote analogues in Early Cambrian archaeocyathan or perhaps algal bioherms. If we turn to localities with soft-bodied animals, we find the strangeness of the mid-Cambrian Burgess Shale fauna that has been strongly emphasized in publications by Whittington (1980) and his associates. Most of the Tommotian fauna appeared inexplicable to its first describers. This marked empirical contrast between the Cambrian and later faunas, quantified by Sepkoski's models, requires the assumption of two phases which need to be explained. The diversity of the 'Cambrian fauna' was measured by Sepkoski (1979, p. 236) as the sum of the familial diversities at successive points in time 'of those classes that are associated with the Cambrian and make substantial contributions to its diversity'. He lists in this category trilobites, monoplacophorans, hyolithids, inarticulate brachiopods, eocrinoids and other echinoderms, Stenothecoidea, 'Calyptomatida' and 'miscellanea' (Volborthellida, Coleolida, Cribricyathea, Mitrosagophora, Hyolithelminthes, Sabelliditida and others). There is obviously a subjective element in this selection but there is no reason why another selection, made by a specialist with different views on classification who may wish to replace some groupings by others, would necessarily produce a significantly different quantitative result. Sepkoski notes that many of the smaller classes die out before the mid-Cambrian peak of diversity while the others 'slowly decline through the rest of the Palaeozoic in almost negative exponential fashion'. In contrast, the remaining 'Palaeozoic fauna diversifies very slowly through most of the Cambrian' and then undergoes an explosive diversification through the Ordovician. It is dominated by articulate brachiopods, the three main classes of Mollusca, crinoids, ostracods, bryozoans, anthozoans, etc. For reasons of space, Sepkoski's comparison of his models with his data (1979, pp. 238–240), which should be studied in detail by interested readers, is here omitted and the validity of his mathematical modelling is accepted. The question of the evolutionary and ecological explanation of his model is relevant to the subject of this chapter.

Some insight may be gained from a consideration of the stochastic aspects of cladogenesis as developed by Gould et al. (1977) but Sepkoski's model suggests additional factors which would be expected to have operated. He relates (1979, p. 246) the two phases to differences between the Cambrian taxa which seem to have been ecologically and adaptively rather 'generalized with broad habitat and trophic requirements', while those producing the diversity of the later 'Palaeozoic fauna' and its dynamic equilibrium state 'appear to have been more specialized with narrower ecologic requirements'. The first phase is that of occupation of a substantially empty ecospace where the first adaptively radiating Metazoa could do things which their unspecialized predecessors could not do (to put the matter in simple words). Sepkoski refers to examples of the generalized habitat requirements of Cambrian faunas consisting – as did their Ediacarian predecessors – largely of surface detritus feeders and grazers and low suspension feeders. He contrasts this with the more varied morphologies and ecologies of Ordovician bivalves, arthropods, echinoderms, etc. and the advancement in community structure, e.g. in the first coral reefs.

The reference to coral reefs is a reminder that an intrinsic evolutionary event such as the advancement in (rugose and tabulate) corals to hermatypic development can change the extrinsic environment by creating new trophic levels and selective advantages leading to new adaptations and specializations. In addition to this example which by its nature is manifest to geological observers, there must have been other instances of interactions of intrinsic and extrinsic factors which can be revealed only by careful palaeontological and ecostratigraphic analysis. Other factors were influencing origination and extinction rates: the first appearance of macro-predators which has yet to be studied in detail, the size increase which is advantageous in certain circumstances though not a general law, the increase in bioenergetic efficiency which must be a general, intrinsic, selective advantage operating with particular force in this unique, initial phase of metazoan expansion. Finally, there is the suspected, if still undocumented, constant factor of increase in biomass in the oceans. If the assumption that the phyto- and zooplankton increased during the Precambrian–Cambrian transition is correct, it would have stimulated, among other steps, the ascent of Cnidaria to the plankton-rich surface level, as documented by the Chondrophora which remained during Palaeozoic time more diversified than in the present seas. According to Stanley's (1973) reasoning, the diversity of both plankton and plankton feeders would have increased simultaneously. In the absence of macro-predators, this would have increased the supply of detritus to benthic detritus feeders, increasing their diversity and abundance as shown by body and trace fossils of Ediacarian and Early Cambrian age. Similar increases would have occurred in suspension feeders. There is a wide field for palaeoecological studies of fossil communities of this age. It is at present the subject of much theorizing but as yet insufficient gathering and analysis of observed facts. It is the merit of the application of ecological theory not only to have drawn attention to this need but also to have made clear the necessary occurrence of interactions between the different kinds of emerging Metazoa and between them and the pre-existing biota.

Evolution is opportunistic. It uses the opportunities created by changes in the physical and biological environment, including those that are due to increasing efficiency of exploitative functions. The biota is a system guided by selection for reproductive advantage of the species. The species reproduces itself, becoming modified either in phyletic lineages or by 'speciation,' i.e. giving rise to new clades. For this reason the results of quantitative population studies (of the ecology of populations of species in ecological time) cannot be applied directly to higher taxa which do not reproduce. We cannot reliably identify as biological species the morphological units in the limited documentary material from the remote geological time of the Precambrian–Cambrian transition with which we operate. If this fundamental difference of levels and scales is recognized and false analogies are avoided, quantitative data on taxonomic diversification in time spans of millions or tens of millions of years can be as valuable and valid as those gathered from observations on evolutionary processes in thousands or tens of thousands of years. They are, however, different kinds of data. The opportunism of evolution does not produce higher taxonomic categories as such. The results of exponential initial diversification of the Metazoa produced not only representatives of phyla as recognized in the classification of the living fauna but also numerous viable organisms of 'uncertain systematic position'. There are no natural groupings of taxa to be labelled '*incertae sedis*' and included in counts of taxonomic diversity as if they were phyla or classes. They are the products of evolutionary experimentation which could be produced and which could flourish in Cambrian times until they were replaced by organisms with more stable, i.e. more efficient biosystems or structural, plans. This phenomenon of evolutionary experimentation is not limited to the Precambrian–Cambrian phase of metazoan diversification. It is still obvious in mid-Cambrian faunas because the special taphonomic conditions of the Burgess Shale provided an uncommonly clear window for observation. A considerable number of 'bizarre' animals '*incertae sedis*' can be seen there. This occurrence does not altogether negate the significance of Sepkoski's exponential curve of Cambrian diversification (Figs. 4.6, 4.7). It involves only a limited rise, a correction which, with more data, could probably be extrapolated over a greater time span without fundamentally changing what the curve tells us.

## Evolutionary diversification

The experimental component of adaptive radiation became more limited in later times, as shown by the Carboniferous Mazon Creek fauna or by the Late Jurassic Solnhofen fauna, both of which contain elements with generally poor preservation potential. Evolutionary diversification in radiation events became channelled in increasingly confined and increasingly familiar lines within otherwise well-known, persistent phyla and classes, but it still produced odd offshoots during phases of rapid diversification. In this phenomenon of evolutionary experimentation at different taxonomic levels lies some explanation of the rapid, temporary rise in apparent rates of evolution at the beginning of Cambrian time. It increased origination rates and, as the experimental and unsuccessful offshoots disappeared again, extinction rates. As far as the origins of the total Phanerozoic diversity are concerned, as far as we are concerned with seeing the dawn of animal life as we know it now, it was evolutionary 'noise' which accompanied the initial explosive (exponential, diversity-dependant) evolution at the time of the Precambrian–Cambrian transition.

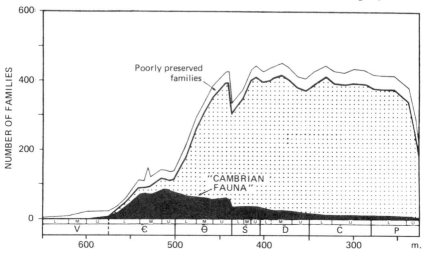

Fig. 4.7. 'The complete stage-by-stage history of familial diversity through the whole of the Palaeozoic, showing "multiple equilibria" with two intervals of logistic diversification. Large letters along the abscissa indicate the geologic systems, smaller letters above indicate the constituent series. Plotting conventions are the same as in Fig. 4[.6], except that total diversities are plotted at the middle of stages during the Palaeozoic equilibrium. The black field in the lower part of the plot represents familial diversity within those classes that are important in or are restricted to the Cambrian... The stippled field above represents the diversity of the remaining skeletonized families, which can be considered as constituting a second "Palaeozoic fauna"' (Sepkoski 1979, caption to Fig. 7).

This was not a critical phase, it was not due to an extraneous event or concatenation of events that altered 'normal' rates of evolution which in any case are unknown or are possibly a global concept of questionable validity. A process which has concurrent stochastic, homeostatic, and selection-activated, adaptive-radiation-inducing components cannot be expected to have also a 'normal' rate – it is inherently more likely to be punctuational (Eldredge & Gould 1972, Stanley 1979) rather than gradational. It occurred because of a temporal coincidence of evolution-enhancing potentialities of these different components, and it occurred as a result of a long sequence of antecedent biohistorical events which were summarized at the end of Chapter 1 as metazoan prehistory. It occurred when metazoan animals had reached a level of ecological competence which enabled them to occupy many previously unoccupied habitats and to adapt to exploitation of their resources.

# 5
# Emerging animal life: thoughts on interactions of lithosphere, hydrosphere, atmosphere and biosphere

## 5.1 Interactions of intrinsic and environmental factors in the emergent biosphere

In the introductory chapter to a recent book on patterns of evolution and the fossil record (Hallam 1977), Stephen J. Gould considered three questions as basic for all past and present palaeontological work. With slight rephrasing, they are: 1. Was the process of life history a directional progressive unfolding or was it a steady-state equilibrium? 2. Was change motivated and determined from the outside, i.e. from the environment, or from within the organism, i.e. internalistic? 3. Was the process of change punctuated by phases of great rapidity or was it a gradual one? The two alternative answers to each of these three questions give eight possible combinations. Gould could fairly easily and clearly categorize the beliefs of the early palaeontologists in terms of these groups. In the concluding chapter of the same book, T. J. M. Schopf classified the viewpoints of the other contributors to it according to the categories which they seemed to fit. It is easy to see that Gould's questions did arise frequently throughout the present study of the particular process of biohistorical change which concerns it. It will also be seen that the answers were hardly ever clearcut. The questions are important and demand answers. They are indeed fundamental in the ongoing debate on the history of life. However, the answers cannot be given by saying yes to one and no to the other alternative offered. The method of alternatives fails to allow for the main factor in the history of life which lies in *interactions*. As if in answer to the first question, Schopf (1980) says in a later book that 'even though the steady-state configuration may seem the most reasonable, perturbations about a "mean" value are sometimes of such enormity that no-steady-state historical factors must be considered' and 'If one consciously supports a steady-state model...then the importance of negative feedback mechanisms becomes of critical importance'. This chapter underlines the truth of these words. Interaction with feedback regulation is the answer to Gould's second question, while the third one partly answers itself here where we are dealing with a 'phase of great rapidity' and want to know how, when and why it happened. Besides, rapid or gradual change in evolution is partly a matter of scale. The amount of 'rapid' morphological change during the Precambrian–Cambrian transition which underlies the exponential increase in taxonomic diversity discussed in the preceding chapter has not been measured as carefully as, for example, G. G. Simpson measured such changes in fossil vertebrates when he coined the term 'quantum evolution'. The reason is the lack of sequences of fossil assemblages representing stages of morphological evolution in Ediacarian time. Since the subject of the present discussion is not evolutionary theory or general patterns of evolution but specifically and exclusively the emergence of metazoan animal life, two questions remain: What were the interactions that led to or resulted from it? Is there any evidence that their rates or intensities increased significantly during the relevant interval of geological time?

We have seen (in Sections 4.6 and 4.7) that intrinsic and environmental factors operated within the biosphere during the historical phase which we are considering and we will recapitulate them. It can hardly be doubted that, as Riedl (1978, p. 266) states, 'all biosystems are selected for continual increase in the throughput of energy'. For the relevant interval of geological time it seems justifiable to assume a steady-state solar energy flux (on which the energy supply to the biosphere ultimately depends). 'By the end of the Precambrian all evidence is consistent with ocean temperatures that were no higher than about 30° C' says Schopf (1980) who assumes that negative feedback mechanisms in the water and the atmosphere were basically responsible for this steady state, even if the solar constant may have varied. The significance of the Late Precambrian glaciations in this context will be considered later (p. 203). If a steady flux of solar energy was in fact available to the emerging animal world, then improvements in the utilization of this energy by animal species, the 'exploitive function' (Hanson 1977, see p. 31 above) must have been a target for selection. The most striking example of such an improvement is the inception of oxidative respiration, which is more than twice as efficient as fermentation. However, this step depended on an environmental factor, the introduction of 'sufficient' oxygen in the atmosphere. Further improvements of efficiency in energy utilization, those which resulted in the evolution of Metazoa from Protista, could be based on physiological and morphological changes and would not have required significant changes in oxygen levels. They occurred while the oxygen level may have changed only from 7–8% to less than 10% PAL, the level assumed by some writers on the subject to have been reached when the land surface became vegetated and suitable for animal life. The intrinsic selection factor of bioenergetic improvement can be considered as one determinant for the increasing diversity of Metazoa, acting during their acoelomate and coelomate radiations. This assumption seems preferable to *ad hoc* hypotheses relying on unproven environmental changes at critical times, in the magnetosphere, atmosphere, hydrosphere (water chemistry), etc. Some of them have already failed tests which have been applied while others have been laid aside pending acquisition of relevant quantitative data for the time interval here considered.

Other intrinsic factors determining evolution are homeostatic in their effects. The evaluation of these homeostatic functions (Hanson 1977) has as much explanatory power for phylogeny as that of other functions. The epigenetic system, the sequence of developmental stages, is essentially conservative. Riedl (1978, p. 258) quotes Waddington as saying 'There are only a certain number of basic patterns which organic form can assume' and is led to a feedback formulation of Haeckel's 'law of recapitulation', based on the 'feedback mechanism produced by the reciprocal effects of phenes and acting back in the organization of the corresponding gene effects'. As Mayr (1963, p. 606) puts it, 'it is...far simpler for the organism to retain...unnecessary aspects of the phenotype than to destroy the harmonious gene complex that controls development'. I take the word 'simpler' to mean less expensive in terms of fitness than any interference with the enormous and still largely unknown complexity of the epigenetic system. It can be lengthened or shortened or re-arranged by selection (Sewertzoff 1931) but its built-in homeostatic resistance to accidental change or total replacement is unquestioned. Hence, early developmental stages of closely related animals tend to resemble each other. To that extent Haeckel's law contains a grain of truth and should not be rejected out of hand. The homeostatically controlled pathways in developmental physiology supply certain structural materials, such as chitin or calcium, at certain levels of evolution. Their use for different purposes produces not only selective advantages but also biomechanical constraints on further evolution. The acquisition of calcareous shells must have been advantageous for many different animals that became capable of using them during the Precambrian–Cambrian transition. Their advantage was controlled by the balance between their function (or functions) and by the biomechanical constraints of the building material such as weight, strength, cost in terms of energy expended on its construction, and effects on locomotion and habitat selection. At some stages in further evolution of some shelly animals, the constraints proved too strong and development of shells was abandoned, reduced, or transferred to the production of minor internal organs, e.g. in coleoid cephalopods or nudibranch gastropods.

The interaction between intrinsic and environmental factors is ubiquitous but neither simple nor straightforward. Its general principles are not well known and not much discussed (see T. J. M. Schopf in Hallam 1977). There is no significant amount of biological theory applicable to that aspect of the early history of the Metazoa. Jägersten's (1972) theory of the evolution of the Metazoa is based on the assumption of a primary pelago-benthic life cycle and its consequences. This cannot be tested palaeontologically since little is known about early zooplankton. The main point arising from a consideration of intrinsic factors is the confirmation of their significance. Profound scepticism is called for when hypotheses are proposed which explain the emergence of the Metazoa by pronounced changes of the environment alone, particularly by invoking isolated physical factors. Riedl (1978) has considered organisms, their development, morphology and evolution from the viewpoint of system and information theory. The organism is a system with its own fourfold pattern: construction ('standard parts'), hierarchy, interdependence, and heredity ('traditive inheritance', Riedl 1978; 'invariance with perturbations', Monod 1970). Its input comes from the physical and the living environment. It manifests itself as the form–function complex (Bock & von Wahlert 1965) or, as Gutmann (1977) puts it: 'The evolving organism is, in every stage of the phylogenetic transformation, conceived of as a network of functional components'. The same must apply to its ontogenetic transformations.

A brief historical reference may be permitted at this point, to illustrate the progress of biological discourse in the last 50 years. The question of whether form or function is dominant in the evolution of animals, hotly debated in the early stages of palaeobiological studies, is seen as irrelevant in more modern approaches. Ultimately, the question of chance and necessity has replaced it, raised by molecular biology and genetics (Monod 1970). Abel (1929) quoted Goethe's dictum 'Thus, form ("Gestalt") determines the animal's mode of life, and its way of life acts back powerfully on all forms' and countered it in 1926 (in German verse, here roughly translated): 'Only the mode of life determines the animal's form; food, locomotion and habitat alone form the body. Any organism must lose in the struggle against forces of its environment unless it adapts, for in the struggle for existence the environment must always win; life is eternally subject to this law'. In recognizing the futility of such dialectic alternatives we have indeed made progress and this gives hope of further progress in the next half-century. At least we shall be able to formulate questions better suited to bring light to the formerly unsuspected complexities of life processes which still tend to be overlooked by unrestrained reductionists (Ayala & Dobzhansky 1974, Mayr 1982).

With the appearance of the Metazoa a new system was introduced in the biosphere. It is organized as tissues and organs among which, in the course of evolution, organs of locomotion and feeding, backed up by sense organs and other life support systems, become repeatedly the most significant innovations. They interact not only with the physical environment but also with the pre- and co-existing biosphere. Metazoa are fitted into trophic levels based on the abundance of primary producers in the form of phytoplankton and its detritus on and in the sediments. We find in Ediacarian deposits evidence of planktotrophic suspension-feeding, detritus-feeding and probably filter-feeding adaptations. Detailed palaeo-ecological studies are still required. We see in these faunas planktonic, nektobenthic and sessile animals and shallow burrowers but we have as yet little knowledge of community relations, particularly in a quantitative sense, and no knowledge of the influence of these faunas on oceanic cycling of chemical elements. Neither morphological nor ethological evidence of macropredators exists for that time. In view of the presence of an abundant food resource for them, this absence or delayed appearance of a class of consumers is an unexpected and significant temporary anomaly. In their absence and in the probable absence of parasites on sessile colonies of coelenterates, the Ediacarian communities had very different structures from those of later times. Earlier suggestions that they must have been atypical for their times, which were possibly made for that reason, have proved unfounded.

Only crude counts of frequency of specimens belonging to different phyla have been made at Ediacara. The clear dominance of medusoid cnidarians and the abundance of sessile colonial coelenterates is seen from the fact that about two-thirds of all fossils found there represents Cnidaria. The Mistaken Point fauna of Newfoundland, which has not yet been systematically described, seems to consist entirely of cnidarians. In the rocks of the Lower Nama Group only a few specimens of fossils other than locally abundant sessile coelenterates and some medusae have been collected, in addition to traces of shallow burrowers. In the richest collections of Ediacarian age, those from the East European Platform, the greatest diversity and abundance is once again seen in coelenterates (Fedonkin 1981a). Counts indicate that among the 500 body fossils collected by 1979, 69% were coelenterates. At Ediacara and in eastern Europe, the next abundant group appears to be the annelid worms. Fedonkin found 25% of his specimens to be 'worms', some of them also known from Ediacara but differently classified by him. Elsewhere only few specimens have so far been recorded from scattered localities but they do generally agree with the established abundance relations.

These community structures, still poorly documented but distinctive, suggest direct exploitation of the lowest trophic level, the living plankton and its dead detritus, as the dominant function of the early metazoans. At this low level of community differentiation, they had little influence on or recognizable interaction with the environment. It is important to note this negative conclusion because it means that the emerging diversity of the Metazoa made its influence on the earlier biota and on the physical environment first felt at a later stage, in the Early Palaeozoic. In fact, we find in the Early Cambrian extensive bioturbation, the first animal bioherms built on a limited scale by Archaeocyatha, and the first evidence of macropredation, for which apparently the highly mobile trilobites were responsible. A peculiar chemical event in the history of oceanic sedimentation which overlaps with the first diversifications of the Metazoa is the appearance in the geological record of marine phosphate deposits. It is not clear in which way these events are related. They add a new pathway in the phosphorus cycle in the sea to the pre-existing ones involving – as far as the biosphere is concerned – only algae and bacteria. The emerging metazoans were, either directly or indirectly, new large-scale consumers of algae and bacteria. Their possible quantitative effect on the phosphorus cycle remains to be examined, as well as the sedimentation of phosphate as fecal matter. This could not have occurred prior to the first appearance of Metazoa with a well-developed digestive system.

The relation between the emerging biota and the atmosphere is the most obvious and the most frequently discussed phenomenon in the history of life. The atmosphere in its present composition is chemically unstable. Existing atmospheres of other planets which differ in composition could not sustain a biosphere like that of the earth. The oxygen in the atmosphere was produced and its amount is kept in a steady state (with possible fluctuations during the last few 100 million years) by interaction with the hydrosphere and the lithosphere (Holland 1978) and with the biosphere (Cloud 1968, 1976a,b, Margulis & Lovelock 1974, 1978). Important for the biosphere as a whole is the availability of nutrients and oxygen. The source of the present oxygen content of the atmosphere is photosynthesis by autotrophic primary producers. Animals, being heterotrophs, rely on green protists and plants for their nutrient supply. It can be assumed to have been present in sufficient quantity where we find fossil animals *in situ*. The oxygen requirements of living animals vary significantly. Facultative aerobic microorganisms can change from fermentation to more efficient oxidative respiration at about 1% PAL. As Cloud (1968) has noted, the oxidative respiration need not have commenced precisely at this 'Pasteur point' of the oxygen accession curve but could have occurred at some other oxygen concentration between 1 and 10% PAL. The oxygen level may have been 3% PAL which according to Cloud is the level required for the metabolism of animals with low oxygen requirements such as annelids. They may have evolved at the time when the level was low, and if that can be confirmed, they could be considered as preadapted to life in present low-oxygen environments.

In his latest work (1978b) Cloud adopts 7% 'at the first appearance of Metazoa' and '10% at the time of the oldest shelled animals'. It is important to note that the present atmospheric level of $O_2$ is 20.95%, hence an increase amounting to about 1.5–2% of oxygen in the atmosphere is considered to have taken place during the Precambrian–Cambrian transition period. It is difficult to see why this should be considered as a significant threshold. At present, animals can exist and respire in conditions where less than one-half of the resulting amount is available.

Another consideration that has played a considerable part in the debate on atmosphere–biosphere interactions is the exclusion of harmful ultraviolet radiation by the ozone in the upper atmosphere. In view of the recent statement that an effective screen of $O_3$ could have existed when $O_2$ was present at 0.1% PAL (Ratner & Walker 1972, Schopf 1980), this event would have occurred long before the first eukaryote cell appeared. Margulis, Walker & Rambler (1976, p. 622) infer from the same data 'that biologically harmful solar ultraviolet radiation was probably not a factor either in the early evolution of eukaryote organisms or in the sudden appearance of abundant Metazoa at the beginning of the Phanerozoic era'. Considering that the evolution of the biosphere had been in progress for some 3000 m.y. before the events of the Precambrian–Cambrian transition, it is perhaps not surprising that its interactions with the environment were as well established as its internal adaptations to the presence of oxygen. No unusual or sudden impetus from outside should be held responsible for the appearance and diversification of the Metazoa. How and when did this process take place?

### 5.2 Metazoan expansion in the marine biosphere: a three-stage process

Sepkoski (1978, 1979) has shown that the 'explosive' diversification of metazoan orders across the Precambrian–Cambrian boundary proceeded at a diversity-dependent, exponential rate. His model appears to describe adequately the observed sequence of events documented in the fossil record. A literal reading of the record starts with the oldest, diversified, well-described and unquestioned assemblage of fossils. Sepkoski, who is interested in the Phanerozoic history, and Cloud (1968, 1978b), who has studied the entire history of life and deleted many erroneous and therefore irrelevant observations from the record, postulate the beginning of animal life in the Early Vendian, about 700 m.y. ago. This reading leaves an apparent gap after the first appearance of eukaryote cells (1300–1500 m.y. ago, Schopf & Oehler 1976). The absence of any fossil record of metazoan life from rocks older than about 1000 m.y. was noted in Chapter 1. Records from younger pre-Vendian (i.e. Upper Riphean) rocks exist but are poorly documented. There are indications of apparent worm burrows and fecal pellets which, if carefully described and analysed, would leave no doubt about the existence of shallow burrowers and sediment feeders 900–1000 m.y. ago. What animals were represented by those traces is a matter of 'metazoan prehistory'. The gap in our knowledge concerning the origination and organization of the first Metazoa can be filled to some extent by biological theory. This was recognized by Cloud (1978b) who favours a polyphyletic origin within a time span of about 100 m.y. The coelomate and the earlier, precoelomate radiations could have occurred in the time interval between 1000 and 850 m.y. ago, or, as Fedonkin (1980c) claims hypothetically, coelomates may even be older. The reason why the pre-Vendian palaeontological record is poor has been discussed before. Its earlier part, the transition from proto- to metazoans occurred, in the opinion of biologists who have discussed it in terms of phylogenetic theories, in microscopically small animals whose chances of preservation in sediments were almost non-existent. The Platyhelminthes which resulted from the acoelomate radiation remain unrecorded as fossils throughout the Phanerozoic era. The absence or extremely rare occurrence of other fossils such as Cnidaria and of trace fossils may not be due only to collection or observation failure but to actual scarcity. This may support the contention that the total biomass, including planktonic, phototrophic, primary producers, was in earlier times significantly smaller than in Palaeozoic time.

In the opinion of Schopf et al. (1973), Metazoa and Metaphyta appeared at about the same time. These authors see the establishment of mitotic-meiotic reproductive mechanisms in eukaryotes as a necessary precondition for their diversification. Stanley (1976b) who agrees with this view, also draws attention to the significance of ecological factors, such as cropping, for the diversity of early Metazoa. All these developments are more likely to have been stepwise than instantaneous. They would have preceded the events of the Precambrian–Cambrian transition. Incidentally, it should be noted that Valentine's (1973*a*, p. 445) five-stage diversification of Metazoa reflects not directly the three-stage process here discussed but the sequence of phylogenetic stages shown in Fig. 3.3 (p. 120). To sum up: despite the paucity or absence of direct evidence from the fossil record, there are good reasons to extend the first phase of existence of Metazoa, at a low level of diversity and abundance, back through the Late Proterozoic Subeon, to about 1000 m.y. ago. On admittedly sketchy theoretical as well as empirical grounds, I consider this interval of about 350 m.y. duration as the first phase, the 'prehistory' or lead time of metazoan diversification.

As was stated before, evolution is opportunistic. If ecological niches suitable for Metazoa were initially unoccupied by organisms of their functional competence, their subsequent occupation would have led to adaptive radiations. They would have included the appearance of many more or less short-lived experimental branches rather than the well-defined phyla which we recognize in the living fauna. Why did the differentiation of these phyla not occur immediately, on the attainment of the metazoan grade of organization? 'When the first Metazoa arose, all types of habitat that could possibly be occupied by Metazoa were unoccupied. The elaboration of metazoan diversity seen in the oldest Phanerozoic rocks (the Ediacarian and Early Cambrian) may thus be interpreted as a function of the many-channelled *adaptive diversification or radiation* of relatively small early metazoan populations' (Cloud 1978*b*, p. 208). The delay of these Early Phanerozoic evolutionary events which correspond to the steep rise in Sepkoski's diversity curve (Fig. 4.1) is due to several necessary factors and precursor events. One is the internal reorganization of the primitive protistan cell colony (or cell colonies of different kinds) to the grade of tissues and organs. This led to the appearance of the first cnidarians, and then to the acoelomate and coelomate radiations. The rates of these evolutionary processes are not known but many steps are required to produce viable organisms at these grades of complexity, ultimately by trial and error. Hence, a considerable span of Proterozoic time would be required before the Vendian level of organismic complexity could be reached.

At about 700–800 m.y. before the present, an external environmental event is known to have occurred which would have slowed down further evolutionary progress. The great Precambrian glaciations occurred during this phase. There is no doubt about their widespread occurrence. They are reported from all continents (except Antarctica). They are the subject of continuing discussions in an ever-growing literature (Crawford & Daily 1971, Kröner & Rankama 1972, Harland & Herod 1975, Williams 1975, Roberts 1976, Kröner 1977, Chumakov 1978, 1981, Frakes 1979, Hambrey

& Harland 1981). There is as yet no agreement about the precise course of events, their causes, and their effects on the biosphere. This last one is the only aspect of the problem that concerns us here. It is made difficult by the obvious stratigraphical and geochronological differences between the Late Precambrian and the Phanerozoic glaciations, which are of limited extent in terms of geographic distribution, stratigraphic thicknesses of rock sequences and duration. Chumakov recognizes three to four separate glacial eras ranging in age from 800–900 m.y. to Vendian, Kröner accepts for Africa two phases between 900 and 600 m.y. ago, while Williams gives a histogram of published ages comprising three groups, from about 955–895, 825–735 and 675–575 m.y. ago. Most known manifestations of Late Precambrian glaciations are of Early Vendian age but some are pre-Vendian, i.e. Late Riphean. The scale of these events makes it clear that they are connected at a global scale but they are unsuited for detailed interregional correlation within time spans of 50–100 m.y. duration. It has not been possible to produce maps of the extent of these glaciations

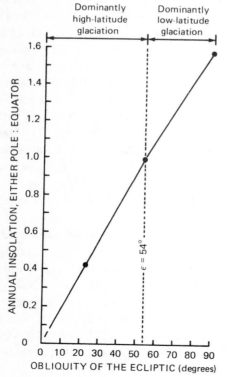

Fig. 5.1. Relation between obliquity of the ecliptic ($\epsilon$) and ratio of annual insolation at either pole to that at the equator. From Williams (1975, Fig. 3), where authorities for data are given.

because maps covering intervals of such duration would make little sense. Besides, climate is related to the physiography of the earth at the relevant time. The positions, outlines and relief of continents and oceans at that time are still uncertain. Indications emerged from early palaeomagnetic measurements that known occurrences of Late Precambrian glacial phenomena originated in low latitudes. The conclusion that this glacial epoch was therefore of worldwide magnitude and extended to within 20–30° of the equator (synchronous global model, W.B. Harland) was soon disputed by Crawford & Daily's (1971) non-synchronous circumpolar model, with continents drifting across polar latitudes. Williams (1975) produced an alternative hypothesis of increased obliquity of the ecliptic (Fig. 5.1), explaining that with the obliquity of the ecliptic increased to 54° from the present 23.5°, low and middle latitudes would be glaciated in preference to the poles. This hypothesis explains a number of other geological observations and avoids results which would have endangered the continued existence of the biosphere, an event which obviously did not occur. New palaeomagnetic data from measurements on Australian rocks led McWilliams & McElhinny (1980) to prefer low latitude for the Late Precambrian glaciation although they noted that each of the various observations leading to this result is not conclusive by itself. The same preference was expressed by Kröner *et al.* (1980) who stated that the present data are not conclusive enough to reject either the low-latitude hypothesis or that of circumpolar glaciation. Further reliable palaeomagnetic data are required to test the Williams hypothesis conclusively and to decide whether or not, as he suggested, the climate at the relevant time was markedly seasonal and climatic zonation was weakened. They are wanted also for the construction of palaeogeographic maps for Late Precambrian–Early Cambrian time. Those produced by Morel & Irving (1978), Ziegler *et al.* (1979) and Zonenshain & Gorodnitsky (1977) were published for discussion rather than as a basis for conclusions about the influence of the distribution of continents and oceans on the climate and the evolution of life. One notes, however, the tendency of the authors to place the continents around the equator. If correct, these configurations would make it possible to account for the distribution of Ediacarian faunas by equatorial seaways and currents. This picture would have to be supported by more conclusive palaeomagnetic measurements before being accepted as the known setting for marine life during the Late Riphean and Early Vendian. None of the contending theories of the Late Precambrian glaciations suggests particularly favourable conditions for the diversification of the Metazoa.

The end of the glaciations and the beginning, in mid-Vendian time, of manifest metazoan diversity marks a lasting change in climate which could be related to the relative abundance of Late Vendian metazoan fossils. Their occurrences postdate the presence of tillites in most of the regions where palaeontological and lithological observations have been made in a clear stratigraphic context (perhaps with the exception of some glacial or cold-climate phenomena in the Lower Schwarzrand Subgroup of the Nama Group, see Kröner & Rankama 1972, Kröner et al. 1980).

The Williams hypothesis of increased obliquity of the ecliptic has so far been neither confirmed nor contradicted by palaeomagnetic data on global low-latitude distribution of glaciated areas. In any case, the long interval of intermittent glaciations between 950 and 570 m.y. ago does not suggest a stimulating influence of fluctuating extreme climatic conditions on the unfolding of metazoan life in the oceans, whatever the physiographic configurations which will eventually be recognized for the various glaciations. On the other hand, the cessation of these profound climatic disturbances, whatever their causes, would undoubtedly have had favourable effects. A time span of 100–150 m.y. duration followed, during which no glaciogenic rocks were formed. In this context the following observation by Williams (1975, p. 458), with reference to the assumed variability of the obliquity of the ecliptic ($\epsilon$), is significant:

> In apparent contrast with late Precambrian glaciations, Late Ordovician continental glaciation was confined largely to *high* palaeolatitudes (Fairbridge 1969, 1970, Beuf et al. 1971). This change in preferred palaeolatitude of glaciation may be explained by a secular decline in $\epsilon$, which caused a relatively rapid reversal or flip-over of climatic zones (when $\simeq 54°$, see Fig. 3) sometime between the latest Precambrian and the Late Ordovician. Perhaps this proposed flip-over of climatic zones accounts also for the explosion of life during the Cambrian.

We are reminded of the second rise of Sepkoski's diversity curve for the 'Palaeozoic fauna' throughout Ordovician time and find his ecological explanation (see p. 191) fitting Williams's postulate of re-established latitudinal climatic zonation and more equable climate when the obliquity of the ecliptic returned to near its present position.

In the overall concept of three stages of metazoan expansion – slow progress during long lead time from inception to mid-Vendian, rapid increase through Cambrian to Late Ordovician, followed by a steady-state plateau throughout the rest of Phanerozoic time – climatic change is seen as a modulating rather than a basic factor. This is in agreement with Schopf's conclusion (1980, p. 247) 'that temperature *per se*, within the known limits in which aquatic life occurs, probably had little to do with *any* pattern of origination or extinction of any group of animal or plant species'. Climatic factors affecting the oceanic biota are linked with bathymetry, circulation, ocean chemistry and ultimately the geographic configuration of oceans and continents, in a complex system of interactions. The evolution of oceanic biota throughout Phanerozoic time is not the subject of the present study. It was examined in detail in Schopf's book which concludes that diversity is a function of habitable area which in turn is a function of number and size of biogeographic provinces, as mediated by sea-level changes. 'Superimposed on the main geological control of biological diversity are "steps" as new adaptive zones are entered, and as diversity increases to local saturation.' The deciphering of the stepwise entering of new adaptive zones in the Ediacarian–Cambrian phase of metazoan diversification will be an important task for future research. It cannot be undertaken now as a general review of the historically unique occupation of previously empty niches in the marine habitat by animals at new levels of organization: most of the basic studies of their palaeoecology have yet to be carried out. Sepkoski's diversity curves for marine orders (1978) and families (1979) show that saturation was reached in Late Ordovician time. This attainment of an equilibrium level between origination and extinction may be related to the first appearance in significant numbers, in the Late Ordovician, of animals at a still higher level of organization, the fishes. This event would have tended to raise the rate of extinction significantly, while the introduction of a new trophic level would have increased diversity, according to the ecological principle of the results of cropping, as discussed for earlier periods by Stanley (1973). This consideration raises the possibility of an alternative explanation of the Late Cambrian 'pseudo-equilibrium', the temporary flattening of the diversity curve observed by Sepkoski. It coincides with a maximum in the development of trilobite diversity. Trilobites may be considered as the first macropredators, but as such they were less efficient than the later fishes. Further palaeontological work will test this hypothesis. The alternative one, of a new climatic regime (according to Williams, as discussed above) will be tested, one hopes, by further palaeomagnetic studies.

### 5.3 Interactions in the light of plate tectonics

The interactions of the biosphere with the atmosphere and hydrosphere in the course of geological time have been mentioned at various points in the present discussion. The relations between the changes in the biosphere with which we are here concerned and those in the lithosphere require further discussion. Biological evolution can have considerable influence on the lithosphere. This occurs directly through increasing deposition of biogenic sediments during geological time, including the formation of skeletal limestones, and of silica from radiolarians and diatoms. It also occurs through sequestration of carbon from organic matter, in bituminous shales and as coal or hydrocarbon deposits, concomitant with the liberation of oxygen to the atmosphere and hydrosphere (discussed by Holland 1978, and by Schopf 1980, with references). Geochemical changes of element distribution between rocks outcropping on the continents, including their plutonic and volcanic rocks, and marine sediments (Veizer in press) may be mediated by biological cycling. The influence on the biosphere of changes in the lithosphere is more relevant here. The extent to which they may throw light on the events during the Precambrian–Cambrian transition, including the emergence of metazoan diversity, deserves further consideration.

The present distribution and diversity of life in the sea depends directly on such factors as depth, currents, temperature, water chemistry, including oxygen and nutrients, which in turn depend on the configuration of ocean basins and continents and their distribution on the surface of the globe. On our knowledge of these changing configurations and distributions depends our understanding of the evolution of the biosphere. The study of observed successive stages of this evolution is the subject of biostratigraphy. With increasing realization of the importance of the interactions discussed in this chapter, efforts are being made to advance biostratigraphy from the level of formal and descriptive cataloguing of these evolutionary changes to a derivation of its principles from palaeogeography, palaeoclimatology and palaeoecology. These advances are slow and will remain slow as long as they are based on the strict application of actualistic principles. The first steps in the earth sciences were based on the dictum that the present is the key to the past. Progress is accelerated with the recognition that actualism has heuristic value but that ultimately it means no more than the immutability of physical laws throughout the universe, through space and time (though 'constants' may have changed).

Palaeogeography was based on the present distribution of continents on the globe until the hypothesis of continental drift was promoted by Wegener (1924, 1929). It was elaborated and made geologically (if not geophysically) more plausible by Du Toit (1937). In the course of the last 20 years the concepts of plate tectonics have incorporated and reframed the hypothesis of continental drift and revolutionized thinking in the earth sciences. The new theory was accepted by the great majority of earth scientists, not because the climate of thought was ripe for a new paradigm but because different sets of measurements could now be made, of global bathymetry, earth magnetism, geochronology of oceanic rocks and geochemistry. They proved to be in general agreement with the theory of ocean floor spreading and movement of lithospheric plates. The new technology of deep-ocean drilling played a vital part in these developments. Biostratigraphy contributed to the deciphering of data from cored ocean sediments. They showed that the ocean floors were young. Their oldest sediments, apart from isolated continental fragments, proved to be of Jurassic age. The continents had moved, not over oceanic crust as Wegener had thought but on rigid plates of lithosphere and upper mantle some 100 km thick. Their configurations and positions could be reconstructed back to Permian times, throughout the last 200 m.y. of earth history. In Permian times the continents were joined in a single continental mass, Pangaea. There is no need to explain here further the plate tectonic theory which is presented in general outlines and in most recent textbooks.

Attempts to follow the past distribution of continents further back to the time of the Precambrian–Cambrian transition have met with considerable difficulties arising from several circumstances. Firstly, the positions of continents can be fixed only by palaeomagnetic measurements from well-dated rocks. These are either lacking for significant areas or are unacceptable because they have been made obsolete by later developments of technical methods. Secondly, there is no direct method of measuring the palaeogeographic longitude of continents. Different assumptions are made by different authors who have made palaeogeographic maps for the Early Palaeozoic. Thirdly, the unravelling of continental collisions along suture lines embedded in the geology of folded mountain zones or former active margins or island arcs presents formidable difficulties and is controversial. Finally, it is thought likely that considerable plate movements occurred during the time of the Precambrian–Cambrian transition.

The process of applying what has been established by deep-sea drilling to the vastly more abundant but more complicated data of continental geology has only started. While it has been said that 'it is well established that the plate tectonic process was operating during the Palaeozoic, and probably in some form during the Proterozoic as well' (Ziegler *et al.* 1979), the process was not necessarily the same as that operating during post-Permian time. The question of former intracratonic versus marginal (collisional) fold belts is the subject of debates. These problems and the evidence for a Proterozoic supercontinent were lucidly discussed by Windley (1977). All authors emphasize the need for further data and their global interpretation. In the meantime it is useful to consider the view on global tectonics in relation to the fossil record of the relevant time span. General principles of the influence of plate movements and consequent configurations of continents and ocean basins on faunas can be applied to or tested against whatever precise palaeogeographic reconstructions emerge from the current debates.

The relevant ecological views expressed by Valentine & Moores (1972) and by Valentine (1973a), summarized here very briefly, indicate that in fluctuating environments the ecosystem is inefficient, few trophic levels can be supported and a low total number of generalized populations can be accommodated. Assembly of continents into a supercontinent is seen as a factor of ecological instability. It tends to enhance marine seasonality. The significance of the still-uncertain latitudinal and longitudinal position of this hypothetical Proterozoic supercontinent must be recognized. However, the general description of climatic, ecological and faunistic conditions appears to fit what is known for Ediacarian time. Particularly striking and seemingly appropriate is the authors' observation that in fluctuating environments short temporary bursts of primary productivity are exploited inefficiently, detritus accumulates, and detritus feeders are accordingly favoured. It was shown that they flourished in Ediacarian faunas.

During times of major transgressions, in the present context during the Early Cambrian transgressions, shelf seas were more widespread and diversity of the neritic fauna increased. Fragmentation of continents tends to isolate shelf areas and this favours speciation. It is increased further by topographic and climatic barriers. The authors link the early radiation of infaunal coelomates with 'unstable conditions associated with continental construction during late Precambrian' (as opposed to later fragmentation). The following Cambrian adaptive radiations suggest the beginning of amelioration of conditions for diversification; the breakup of the (still hypothetical) Proterozoic supercontinent favours high diversity. There is increasing stability of trophic resources; wide shelves related to transgressions improve climatic stability and create varied habitats. They lead to further isolation of populations, increased provinciality and partitioning of environments. These factors favour increased diversity. The authors conclude: 'High diversities correlate with stable and low diversities with fluctuating regimes... Precambrian and Cambrian radiations from which the higher invertebrate phyla developed may have been primarily an adaptive response to changes in trophic resource regimes brought about by changing land–sea patterns'. These patterns – size, height of continental surfaces during transgressions and relief (insofar as trends of major mountain ranges affect rainfall, also their latitudinal and longitudinal distribution) – mediate climate. This in turn acts through food and solar energy resources on the biosphere. These leads from ecological theory to evolutionary palaeobiology can be followed only when the development of plate tectonics for the Late Precambrian and Early Cambrian permits the construction of reliable palaeogeographic maps. They should show not only plate boundaries but the developing transgressions and regressions, climatic zones and palaeo-oceanography. All this is still in the future but a lead in this direction was given by Ziegler *et al.* (1979) for Palaeozoic time. Valentine (1973*a*) refers to the 'interplay' between evolutionary trends and environmental fluctuations. The interactions of lithosphere and biosphere are definitely on the agenda for further critical evaluation.

## 5.4 Animal life: past, present and future

The discussion of the course taken by the emerging animal life and of the timing of steps in the attainment of its diversity has led to the conclusion that a fuller and deeper understanding of the complex interactions between the lithosphere, hydrosphere, atmosphere and biosphere is necessary for further elucidation of the process. Increasing numbers of recent studies of the history of life on earth mention the need for a widening of the scope of these investigations. The need to integrate them with the history of the earth's crust and of the atmosphere, particularly with the data of palaeo-oceanography, palaeogeography and palaeoclimatology, is as apparent to students of historical biology as it is to specialists in the other disciplines. This notion accords with the present trend towards interdisciplinary studies. Current strategy in dealing with newly discovered or previously misinterpreted phenomena, such as those which occurred during the Precambrian–Cambrian transition, is not altogether satisfactory. It results often in proposals of *ad hoc* hypotheses for each group of new facts. Examples have been cited but there are many others. Data from other fields of studies have been used in explanations, either in isolation or in heuristic combinations: atmospheric physics, cosmology, genetics, geophysics, geochemistry, ecological theory, have all been used for the construction of specific models for isolated groups of data. They can be useful, as far as they are testable and linked with other relevant observations. Progress in the understanding of the historical process is delayed when events are mistimed through neglect of geochronology or its misapplication, or when they are treated as singularities or 'crises', or else grouped as cycles, without sufficient evidence for one or the other of these alternatives. Cross-connections of synchronous events may be more significant than either separate explanations for each, or their indiscriminate fitting into a preconceived view of the historical process.

Actualistic or uniformitarian approaches to processes in earth history can be and have been misapplied, simply because not everything that happened in history is still happening and observable now. The concept of historical biology as a field of scientific research seems nonsensical to some biologists who assert that life cannot be studied by observing dead matter, without experimental verification. Yet, few if any biologists will deny that life had a scientifically significant history. Its main course is known as a sequence of events in chronological order. We know the approximate duration of the long periods of existence of Monera (Prokaryota), of Protista, and we have discussed the emergence of Animalia, its slow initial phase or 'prehistory', and the following spectacular exponential diversification, up to a point in time when some kind of equilibrium between origination and extinction was reached. We have seen reasons to expect that the study of interactions between this emergent biosphere and the atmosphere, hydrosphere and lithosphere will lead to a fuller understanding of the processes which are involved, and their rates. It may be useful to extend, in a very sketchy manner, the consideration of the history of the kingdom Animalia to the present so that we can see whether we can draw some tentative conclusions about the future. This, after all, is said to be the aim of all good science.

The major event in the biosphere in the course of Phanerozoic time was the emergence on land, some 400 m.y. ago, of vascular plants (Tracheophyta) and terrestrial Metazoa. The ecological coincidence of these events led to two evolutionary innovations of great significance for the present state of the biosphere. The vascular plants, being able to exploit sunlight much more efficiently than in the confines of their previous environment, flourished, occupying and creating new ecospace; the insects among the arthropods became the numerically dominant animals. The vertebrates were set on their evolutionary pathways that would bring them eventually to the exploitatively and reproductively most efficient level, that of the mammals, including man. The Phanerozoic history of the marine and terrestrial biota has been described and analysed many times and in much detail. It needs no further comment here except for a few points which confirm or add information to what we have found in retracing the history of the Metazoa towards its Precambrian beginnings.

(1) The diversity equilibrium of the Metazoa went through several critical phases of accelerated extinction. Their causes are still disputed. Two of them mark, respectively, the ends of the Palaeozoic and Mesozoic Eras. When examined in detail, the more spectacular extinctions are seen to be preceded by periods of decline of varying lengths (graptolites, trilobites, ammonites, which came close to extinction twice before). This observation and the fact that many contemporaneous taxa remained unaccountably unaffected by otherwise widespread extinctions makes it difficult to accept catastrophic global or extraterrestrial causes. Trophic resource limitation and climatic change seem most likely factors.

(2) Several adaptive radiations of major taxa occurred after long lead times lasting from tens to a hundred million years, e.g. in Mammalia.

This means that the lead time here proposed for the early history of the Eukaryota before the Phanerozoic burst of diversification is not uncommon, and slow elaboration of new ground plans by trial and error, with many unsuccessful side branches, may be the rule rather than the exception.

(3) The Permian glacial period which affected only the then circumpolar Gondwana continents, and the Pleistocene glaciations which may not have ended yet, had only limited rather than dramatic biological effects, mainly through oceanographic and other physiographic changes. Neither period is comparable in duration and apparent complexity of phases to the Late Precambrian glacial periods. Obviously, detailed knowledge of global physiography is required before the influence of palaeoclimate on evolution can be usefully discussed on a historical rather than theoretical basis.

(4) Most possible body plans of Metazoa seem to have evolved and most existing ecological niches seem to have been occupied during Phanerozoic time. The first of these propositions is often inaccurately expressed by the dictum that all existing phyla appeared in the early Palaeozoic.

The ancient idea of a 'ladder of life' leading to the species *Homo sapiens* on the highest rung has still some influence on our thinking but it is expressed in less blatantly anthropocentric terms. The fact that arthropods are the most abundant and have the most varied adapted forms of animal life gives reasons for questioning the domination of the biosphere by man, at least in a strictly biological sense. However, there is no doubt about the significance of the evolutionary emergence of *Homo sapiens* for the fleeting present and the indefinite future of the biosphere and of animal life in it. What sets us apart as a peculiar biological species is a unique direction taken in its evolution. The essence of heredity is seen in modern biology as the transmission of information from one generation to the next. Evolution is the selective modification of this information which is contained in the genetic code. In the course of evolution of the genus *Homo* and its species during the last 4 m.y., an entirely new system of transmission of information has been added. This is based on the development of language, concomitant with that of abstract thought, introspection, and other higher brain functions. Normal morphological, physiological and biomechanical evolution has not necessarily been superseded by this development but its rates are slow compared with the explosive development of the higher nervous system ('mind'). Many human beings complain that effects of morphological evolution were too slow for comfort: the upright posture which evolution forced on our mammalian body plan 4 m.y. ago is still maladjusted, as evidenced by circulatory, orthopaedic, respiratory and dental troubles from which many of us suffer. The difference between the transmission of information in humans and in other vertebrates is fundamental and its consequences are great. The population explosion in the species *H. sapiens* is a unique event, perhaps not in numbers, but certainly in its impact on the trophic and energy resources on which its continued existence depends (for a penetrating and deeply humanistic discussion of these problems, see Cloud 1978*b*, Chapters 19 and 20). Its impact extends to the atmosphere and hydrosphere in a manner that appears to threaten the continued existence of the human and other species.

Our concern here is primarily with the continued existence of the animals and the biosphere as a whole. Assuming that the higher mental functions, in totality called intelligence, have not reached the necessary level at which mankind can heed the warnings of thinkers and mentors, population pressure and depletion of resources may lead to its extinction. The evolutionary potential remaining in the residual, perhaps already damaged, biosphere is such that it could take off again at some other point and evolve in new directions. We did not find the course of evolution of the Metazoa to be a unidirectional climb up the ladder of life from Protozoa to man. The fossil record tells of different sequences of events. The shell-bearing Protista evolved and flourished long after the 'age of protists' had ended. The different multicellular descendants of different protists may have appeared at different times, well into the Phanerozoic. New building plans of metazoan grades will appear wherever there are unoccupied ecological niches requiring new adaptations. The tolerance of life forms to extreme physical and chemical conditions is enormous but at present they are utilized only within the more favourable ranges of their parameters by significant numbers of organisms. In a perceptive classical work, the biochemist L. J. Henderson (1913) wondered about the fitness of the environment, meaning essentially the physicochemical properties of the chemical elements without which life could not exist. It is now considered that it does not exist on other planets with different distributions of elements. It does not seem likely, at this stage at least, that the excessive proliferation of one species can destroy the physicochemical basis of life itself. The interactions will not cease between a biosphere built on this basis, an atmosphere interacting with it, a hydrosphere, or a lithosphere which can be depleted of resources for and by human consumption but not of the essentials for life as such. The five-kingdom concept tells us that the organic world consists of the ubiquitous, rapidly proliferating and metabolically almost infinitely versatile Monera, the Protista, some of which can live either as animals or as plants, the saprophytic Fungi, the autotrophic plants and the heterotrophic animals. No end of these five kingdoms is in sight. Their resources are not threatened and are renewable.

The interactions between the atmosphere and the biosphere, in particular, have been the subject of a recent book by Lovelock (1979), an atmospheric scientist, whose previous joint papers with Margulis (Lovelock & Margulis 1974, Margulis & Lovelock 1974, 1978) should also be noted. They concern mostly the composition of the atmosphere and its modification and modulation by microbiotas. For that reason they have not been previously considered in this study. However, these publications suggest a 'Gaia hypothesis' according to which the composition of the atmosphere 'with respect to reactive gases, the acidity and the temperature of the earth's lower atmosphere are maintained near optima *by the biota for the biota*' (Margulis & Lovelock 1978, p. 243, my italics). Lovelock goes further and considers the earth with its atmosphere and hydrosphere as a kind of superorganism with its inherent homeostatic controls. I venture to suggest that the 'Gaia hypothesis' is not so much explicative as normative. In a timely and impassioned statement it tells people what to do and what not to do if the earth systems are to continue to function homeostatically, for their own benefit and for that of other life forms. Holland (1978, p. 317) believes that the close chemical links between atmosphere, hydrosphere and biosphere 'do not imply that the biosphere is an adaptive control system capable of maintaining the earth in homeostasis...The ocean–atmosphere system has responded to biologic inventions such as photosynthesis by adjusting its composition so that the bulk chemistry of the crust is maintained. The biosphere has responded to these adjustments by optimizing the use of available free energy. The long-term stability of the system is due to the interplay between the numerous inorganic and biologic processes that have shaped the surface of the earth.' There is no sign of an end of this interplay, these interactions, as long as there is a sun and an earth, and the laws of thermodynamics remain valid. Whether man remains to watch, enjoy and interpret this interplay depends on whether he learns in time to treat this complex system with the care that the most fragile of organisms would need and deserve, in order to survive.

We have not searched for a vestige of the beginning. We have seen something of the dawn of animal life. We have used the light of this day to study it. There is, we repeat thankfully, no prospect of an end.

# Appendix: Questions of language and terminology

(1) *Languages*. Quotations from Russian publications are given in my own translation. Spelling of English words follows generally the preferred versions of the Concise Oxford Dictionary, 6th edn.

(2) *Geographic terminology*. Geographic terms are used without political implications. The descriptive term 'southwest Africa' is being replaced in official publications by 'Southwest Africa/Namibia'. Both are used in the text. Further change cannot be anticipated.

(3) *Stratigraphic terminology*. (*a*) Russian terms. Russian stratigraphic classifications use different sets of terms with meanings which do not invariably coincide with English terms. A *Stratigraphic code of the USSR* was published in Russian with complete English translation (Zhamoida 1979) but it does not overcome the divergence of basic concepts nor the difficulty of transliteration of geographic parts of unit names which often requires knowledge of the names of small villages or rivers from which they are derived. The 'Rules of derivation and orthography of names of stratigraphic units' in the translated code are of limited assistance. Full uniformity of names used in this text has not been achieved. One important stratigraphic term in the Russian literature is 'Horizon'. It is defined as a regional unit determined (for Phanerozoic stratigraphy) mainly by palaeontological criteria (Zhamoida 1979, pp. 24, 93). The translation 'stratohorizon' is offered, while 'regional stage' is not recommended. A. I. Zhamoida (personal communication) explained that it is a regional generalization of local suites and series (approximately formations and groups in Anglo-American usage) for entire sedimentary basins. (*b*) A name for the Early Vendian. Asklund (1958) named the sequence of strata from the base of the Scandinavian tillites to the base of the fossiliferous Cambrian the 'Varegian' (a chronostratigraphic term in Störmer's opinion), explaining that this name refers to the discovery of tillites on the

# Appendix: Questions of language and terminology

Varanger Peninsula but equally to the Varegians, a tribal name given by the ancient Greeks to the Vikings. He invoked the parallel with the formation of the names Cambrian, Ordovician and Silurian from ancient tribal names. This double derivation has led to confusion. Harland & Herod (1975) restricted the term by separating the Ediacarian from the preceding tillite-bearing 'Varangian'. This hybrid of the geographic and ethnographic roots of Asklund's term is now undergoing a change to 'Varangerian', of purely geographic derivation. Uniformity has not been achieved in this text.

(4) *Biological terminology*. The term 'blue-green algae' and its equivalent 'Cyanophyta' are apparently now in the process of being changed to 'Cyanobacteria'. There are good reasons for, as well as objections against, this change. Unable to foresee the outcome of these arguments, I have not achieved uniformity in this text.

# References

(Titles of Russian papers published without English titles are given in English translation, in parenthesis. Russian original publications have been used in preference to published summaries or translations because of their occasional imperfections.)

Abel, O. (1929). *Paläobiologie und Stammesgeschichte*. G. Fischer, Jena.
Alpert, S. P. (1977). Trace fossils and the basal Cambrian boundary. *Geol. J.*, Spec. Issue 9, 1–8.
Anderson, M. M. (1972). A possible time span for the Late Precambrian of the Avalon Peninsula, southeastern Newfoundland, in the light of world-wide correlation of fossils, tillites, and rock units within the succession. *Canadian J. Earth Sci.*, 9, 1710–26.
Anderson, M. M. (1978). Ediacaran fauna. In: *McGraw Hill yearbook of science and technol*, pp. 146–9. McGraw Hill, New York.
Anderson, M. M. & Misra, S. B. (1968). Fossils found in the Pre-Cambrian Conception Group of south-eastern Newfoundland. *Nature*, 220, 680–1.
Asklund, B. (1958). Le problème Cambrien–Eocambrien dans la partie centrale des Calédonides suédoises. *Centre Nat. Rech. Sci. Colloques Int.*, 76, 39–52.
Avnimelech, M. (1955). Occurrence of fossil Phoronidea-like tubes in several geological formations in Israel. *Bull. Res. Council Israel*, 58, 174–7.
Awramik, S. M. (1971). Precambrian stromatolite diversity: reflection of metazoan appearance. *Science*, 174, 825–7.
Awramik, S. M. & Barghoorn, E. S. (1977). The Gunflint microbiota. *Precambrian Res.*, 5, 121–42.
Ayala, F. J. & Dobzhansky, Th. (1974). *Studies in the philosophy of biology, reduction and related problems*. Macmillan, London.
Banks, N. L. (1970). Trace fossils from the Precambrian and Lower Cambrian of Finnmark, Norway. *Geol. J.*, Spec. Issue 3, 19–34.
Banks, N. L. (1973). Trace fossils in the Halkkavarre section of the Dividal Group (?Late Precambrian–Lower Cambrian), Finnmark. *Norsk Geol. Unders.* 269, 197–236.
Barghoorn, E. S. & Schopf, J. W. (1966). Microorganisms three billion years old from the Precambrian of South Africa. *Science*, 152, 758–63.
Barnes, R. D. (1980). *Invertebrate zoology* (4th edn). Saunders College, Philadelphia.
Beckinsale, R. D., Thorpe, R. S., Pankhurst, R. J. & Evans, J. A. (1981). Rb–Sr whole rock isochron evidence for the age of the Malvern Hills complex. *J. Geol. Soc. Lond.*, 138, 69–73.
Beer, E. J. (1919). Note on a spiral impression on Lower Vindhyan limestone. *Rec. Geol. Survey India*, 50, 139.

de Beer, G. N. (1951). *Embryos and ancestors* (2nd edn). Clarendon Press, Oxford.
Bekker, Yu. R. (1977). (The first palaeontological finds in the Riphean of the Ural.) *Izv. Akad. Nauk SSSR*, Ser. Geol., No. 3, 90–100. (In Russian.)
Beklemishev, V. N. (1964, in Russian, 3rd edn., 1969, English transl.). *Principles of comparative anatomy of invertebrates.* 2 vols. Univ. Chicago Press.
Bengtson, S. (1970). The Lower Cambrian fossil *Tommotia. Lethaia*, **3**, 363–92.
Bengtson, S. (1976). The structure of some Middle Cambrian conodonts, and the early evolution of conodont structure and function. *Lethaia*, **9**, 185–206.
Bengtson, S. (1977a). Aspects of problematic fossils in the Early Palaeozoic. *Acta Univ. Upsaliensis*, **415**, 71 pp.
Bengtson, S. (1977b) Early Cambrian button-shaped phosphatic microfossils from the Siberian Platform. *Palaeontology*, **20**, 751–62.
Bengtson, S. (1978). The Machaeridia – a square peg in a pentagonal hole. *Thalassia Jugoslavica*, **12**, 1–10.
Bengtson, S. & Missarzhevsky, V. V. (1981). Coeloscleritophora – a major group of enigmatic Cambrian metazoans. *Short Pap. 2nd Int. Symp. Cambrian Syst.*, Open-File Rep. 81–743, 19–21. (United States Geological Survey.)
Berkner, L. V. & Marshall, L. C. (1965). History of major atmospheric components. *Proc. Nat. Acad. Sci.*, **53**, 1215–25.
Bernal, J. D. (1951). *The physical basis of life.* Routledge & Kegan Paul, London.
Bernal, J. D. (1967). *The origin of life.* Weidenfeld & Nicholson, London.
Beuf, S., Biju-Duval, B., de Charpal, O., Rognon, P., Gariel, O. & Bennacef, A. (1971). *Les grès du Paléozoique inférieur au Sahara.* Technip, Paris.
Beurlen, K. & Sommer, F. W. (1957). *Observaçoes estratigráficas e paleontológicas sobre o calcário Corumbá. D.N.P.M., D.G.M.*, Bol. 168. Rio de Janeiro. 35 pp.
Binda, P. L. & Bokhari, M. M. (1980). Chitinozoanlike microfossils in a Late Precambrian dolostone from Saudi Arabia. *Geology*, **8**, 70–1.
Birket-Smith, S. J. R. (1981a). A reconstruction of the Pre-Cambrian *Spriggina. Zool. Jb. Anat.*, **105**, 237–58.
Birket-Smith, S. J. R. (1981b). Is *Praecambridium* a juvenile *Spriggina? Zool Jb. Anat.*, **106**, 233–5.
Bischoff, G. C. O. (1978a). *Septodaeum siluricum*, a representative of a new subclass Septodaearia of the Anthozoa, with partial preservation of soft parts. *Senckenbergiana Lethaea*, **59**, 229–73.
Bischoff, G. C. O. (1978b). Internal structures of conulariid tests and their functional significance, with special reference to Circonulariina, n. suborder (Cnidaria, Scyphozoa). *Senckenbergiana Lethaea*, **59**, 275–327.
Bloeser, B., Schopf, J. W., Horodyski, R. J. & Breed, W. J. (1977). Chitinozans from the Late Precambrian of the Grand Canyon, Arizona. *Science*, **195**, 676–9.
Boaden, P. J. S. (1975). Anaerobiosis, meiofauna and early metazoan evolution. *Zool. Scripta*, **4**, 21–4.
Bock, W. J. & von Wahlert, G. (1965). Adaptation and the form–function complex. *Evolution*, **19**, 269–99.
Borello, A. (1966). Trazas, restos tubiformes y cuerpos fósiles problemáticos de la Fm. La Tinta, Sierras Septemtrionales, Pcia de Buenos Aires. *Paleontogr. Bonaerense*, **5** (Com. Inv. Cient. Pcia Buenos Aires), 1–42.
Borovikov, L. I. (1976). (The first find of fossil remains of *Dickinsonia* in the Lower Cambrian deposits of the USSR.) *Doklady Akad. Nauk SSSR*, **231**, 1182–4. (In Russian.)
Bouillon, J. (1968). Introduction to Coelenterates. In: M. Florkin & B. J. Scheer (eds.), *Chemical zoology*, vol. 2, pp. 81–147. Academic Press, New York.
Boyden, A. (1947). Homology and analogy. *Am. Midl. Nat.*, **37**, 648–69.
Brasier, M. D. (1979). The Cambrian radiation event. In: M. R. House (ed.), *The origin of major invertebrate groups*, pp. 103–59. Academic Press, London.

Brasier, M. D. (1980). The Lower Cambrian transgression and glauconite–phosphate facies in western Europe. *J. Geol. Soc. Lond.* **137**, 695–703.

Brasier, M. D. (1982). Sea-level changes, facies changes and the Late Precambrian–Early Cambrian evolutionary explosion. *Precambrian Res.*, **17**, 105–23.

Brasier, M. D. & Hewitt, R. A. (1979). Environmental setting of fossiliferous rocks from the uppermost Proterozoic–Lower Cambrian of central England. *Palaeogeogr., Palaeoclimat., Palaeoecol.*, **27**, 35–57.

Brasier, M. D. & Hewitt, R. A. (1981). Faunal sequence within the Lower Cambrian 'non-trilobite' zone (s.l.) of central England and correlated regions. *Short Pap. 2nd Int. Symp. Cambrian Syst.*, Open-File Rep. 81–743, 29–33. (United States Geological Survey.)

Brasier, M. D., Hewitt, R. A. & Brasier, J. (1978). On the Late Precambrian–Early Cambrian Hartshill Formation of Warwickshire. *Geol. Mag.*, **115**, 21–36.

Brien, P. (1969). *Principes de phylogénèse animale appliqués à l'interprétation des métazoaires diploblastiques et à leur évolution en métazoaires triploblastiques. Acad. Roy. Belg. Cl. des Sci. Coll. 8°*, 2e Sér., vol. 38, fasc. 7, 110 pp.

Broda, E. (1978). *The evolution of bioenergetic processes* (revised reprint). Pergamon Press, Oxford.

Calvin, M. (1969). *Chemical evolution*. Clarendon Press, Oxford.

Cavalier-Smith, T. (1975). The origin of nuclei and of eucaryotic cells. *Nature*, **256**, 463–8.

Chang, S. (1981). Organic chemical evolution. In: J. Billingham (ed.), *Life in the universe*, pp. 21–46. MIT Press, Cambridge, Mass.

Chumakov, N. M. (1978). (*Precambrian tillites and tilloids.*) 'Nauka', Moscow. (In Russian.)

Chumakov, N. M. (1981). Upper Proterozoic glaciogenic rocks and their stratigraphic significance. *Precambrian Res.* **15**, 373–95.

Chumakov, N. M. & Semikhatov, M. A. (1981). Riphean and Vendian of the USSR. *Precambrian Res.*, **15**, 229–53.

Chung Fu-Tao (1977). On the Sinian geochronological scale of China, based on isotopic ages for the Sinian strata in the Yenshan region, north China. *Sci. Sinica*, **20**, 818–34.

Cisne, J. L. (1974). Trilobites and the origin of arthropods. *Science*, **186**, 13–18.

Clark, R. B. (1964). *Dynamics in metazoan evolution, the origin of the coelom and segments*. Clarendon Press, Oxford.

Clark, R. B. (1969). Systematics and phylogeny: Annelida, Echiura, Sipunculida. In: M. Florkin & B. T. Scheer (eds.), *Chemical zoology*, vol. **4**, pp. 1–68. Academic Press, London.

Clark, R. B. (1979). Radiation of the Metazoa. In: House, M. R. (ed.), *The origin of major invertebrate groups*, pp. 55–102. Academic Press, London.

Clemmey, H. (1976). World's oldest animal traces. *Nature*, **261**, 576–8.

Cloud, P. E. (1968). Pre-metazoan evolution and the origins of the Metazoa. In: T. Drake (ed.), *Evolution and environment*, pp. 1–72. Yale Univ. Press, New Haven.

Cloud, P. (1973). Pseudofossils: a plea for caution. *Geology*, **1**, 123–7.

Cloud, P. (1976a). Beginnings of biospheric evolution and their biogeochemical consequences. *Paleobiology*, **2**, 351–87.

Cloud, P. (1976b). Major features of crustal evolution. *Trans. Geol. Soc. S. Africa*, Annex to vol. 79 (Alex L. Du Toit Memorial Lecture No. 14), 1–33.

Cloud, P. (1978a). World's oldest animal traces. *Nature*, **275**, 344. (With reply by H. Clemmey, pp. 344–5.)

Cloud, P. (1978b). *Cosmos, earth and man. A short history of the universe*. Yale Univ. Press, New Haven.

Cloud, P. & Glaessner, M. F. (1982). The Ediacarian Period and System: Metazoa inherit the earth. *Science*, **217**, 783–92.

Cloud, P., Gustafson, L. B. & Watson, J. A. L. (1980). The works of living social insects as pseudofossils and the age of the oldest known Metazoa. *Science*, **210**, 1013–15.

Cloud, P., Wright, J. & Glover, L. (1976). Traces of animal life from 620-million-year-old rocks in North Carolina. *Am. Scient.*, **64**, 396–406.

Coats, R. P. (1973). *Explanatory notes, 1:250,000 Geological Series Sheet SH/54-9*. Geol. Survey S. Australia.

Coats, R. P. & Preiss, W. V. (1980). Stratigraphic and geochronological reinterpretation of Late Proterozoic glaciogenic sequences in the Kimberley region, Western Australia. *Precambrian Res.*, **13**, 181–208.

Cohee, G. V., Glaessner, M. F. & Hedberg, H. D. (eds.) (1978). *Contributions to the geologic time scale*. Am. Ass. Petroleum Geol., Studies in Geology, No. 6.

Conway Morris, S. (1976). A new Cambrian lophophorate from the Burgess Shale of British Columbia. *Palaeontology*, **19**, 199–222.

Conway Morris, S. (1977). Fossil priapulid worms. *Spec. Pap. Palaeont.*, **20**, 1–95.

Conway Morris, S. (1979a). Middle Cambrian polychaetes from the Burgess Shale of British Columbia. *Phil. Trans. Roy. Soc. Lond.*, B, **285**, 227–74.

Conway Morris, S. (1979b). The Burgess Shale (Middle Cambrian) fauna. *Ann. Rev. Ecol. Syst.*, **10**, 327–49.

Conway Morris, S. (1981). Parasites and the fossil record. *Parasitology*, **82**, 489–509.

Conway Morris, S. (1982). *Wiwaxia corrugata* (Matthew), a problematical Middle Cambrian animal from the Burgess Shale of British Columbia. In: *Third North American palaeontological Convention, proc.*, vol. 1, pp. 93–8.

Conway Morris, S. & Crompton, D. W. T. (1982). The origin and evolution of the Acanthocephala. *Biol. Rev.*, **57**, 85–115.

Conway Morris, S. & Fritz, W. H. (1980). Shelly microfossils near the Precambrian–Cambrian boundary, Mackenzie Mountains, northwestern Canada. *Nature*, **286**, 381–4.

Cope, J. C. W. (1977). An Ediacara-type fauna from South Wales. *Nature*, **268**, 624.

Cowie, J. W. (1981). The Proterozoic–Phanerozoic transition and the Precambrian–Cambrian boundary. *Precambrian Res.*, **15**, 199–206.

Cowie, J. W. & Glaessner, M. F. (1975). The Precambrian–Cambrian boundary: a Symposium. *Earth-Sci. Rev.*, **11**, 209–51.

Crawford, A. R. & Daily, B. (1971). Probable non-synchroneity of Late Precambrian glaciations. *Nature*, **230**, 111–12.

Cribb, S. J. (1975). Rubidium–strontium ages and strontium isotope ratios from the igneous rocks of Leicestershire. *J. Geol. Soc. Lond.*, **131**, 203–12.

Crimes, T. P., Legg, I., Marcos, A. & Arboleya, M. (1977). ?Late Precambrian–low Lower Cambrian trace fossils from Spain. *Geol. J.*, Spec. Issue 9, 91–138.

Crimes, T. P. & Germs, G. J. B. (1982). Trace fossils from the Nama Group (Precambrian–Cambrian) of southwest Africa (Namibia). *J. Paleont.* **56**, 890–907.

Daily, B. (1972). The base of the Cambrian and the first Cambrian faunas. *Centre for Precambrian Res., Univ. Adelaide* Spec. Pap., 1, 13–37.

Daily, B. (1973). Discovery and significance of basal Cambrian Uratanna Formation, Mt Scott Range, Flinders Ranges, South Australia. *Search*, **4**, 202–5.

Daily, B. (1976). (New data on the base of the Cambrian in South Australia.) *Izv. Akad. Nauk SSSR* Ser. Geol., No. 3, 45–51. (In Russian.)

Dalgarno, C. R. (1964). Report on the Lower Cambrian stratigraphy of the Flinders Ranges, South Australia. *Trans. Roy. Soc. S. Aust.*, **88**, 129–44.

Dalla Salda, L. (1979). Nama and La Tinta Group – a common South Africa–Argentine basin? *Chamber of Mines Precamb. Res. Unit, Dept. Geol., Univ. Cape Town*, 16th Ann. Rep., 113–28.

Debrenne, F. & James, N. P. (1981). Reef-associated archaeocyathans from the Lower Cambrian of Labrador and Newfoundland. *Palaeontology*, **24**, 343–78.

Degens, E. T., Johanesson, B. W. & Meyer, R. W. (1967). Mineralization processes in molluscs and their paleontological significance. *Naturwissensch.*, **54**, 638–40.

Derstler, K. (1981). Morphological diversity of Early Cambrian echinoderms. *Short pap. 2nd Int. Symp. Cambrian Syst.*, Open-File Rep. 81–743, 71–5. (United States Geological Survey.)

Ding Qixiu & Chen Yiyuan (1981). Discovery of soft metazoan from the Sinian System along eastern Yangtse Gorge, Hubei. *J. Wuhan College Geol.*, No. 2, 53–7.

Dougherty, E. (ed.) (1963). *The lower Metazoa. Comparative biology and phylogeny*. Univ. Calif. Press, Berkeley.

Downie, C., Evitt, W. R. & Sarjeant, W. A. S. (1963). Dinoflagellata, Hystrichosphaeridia and the class of Acritarcha. *Stanford Univ. Publ. Geol. Sci.*, **7**, 3–16.

Dunning, F. W. (1976). Central England and Welsh Borders. *Spec. Rep. Geol. Soc. Lond.*, No. 6, 83–95.

Durham, J. W. (1978). The probable metazoan biota of the Precambrian as indicated by the subsequent record. *Ann. Rev. Earth Planet. Sci.*, **6**, 21–42.

Du Toit, A. L. (1937). *Our wandering continents: an hypothesis of continental drifting*. Oliver & Boyd, London.

Eldredge, N. & Gould, S. J. (1972). Punctuated equilibria: an alternative to phyletic gradualism. In: T. J. M. Schopf (ed.), *Models in paleobiology*, pp. 82–115. Freeman, Cooper & Co., San Francisco.

Evitt, W. R. (1963). A discussion and proposals concerning fossil dinoflagellates, hystrichospheres, and acritarchs, II. *Proc. Nat. Acad. Sci.*, **49**, 298–302.

Fairbridge, R. W. (1969). Early Palaeozoic south pole in northwest Africa. *Bull. Geol Soc. Am.*, **80**, 113–14.

Fairbridge, R. W. (1970). South pole reaches the Sahara. *Science*, **168**, 878–81.

Fairchild, T. R. (1978). Evidências paleontologicas de una possivel idade 'Ediacariana' ou Cambriana inferior, para parte de Grupo Corumbá (Mato Grosso do Sul). *Soc. Brasil. de Geol. XXX Congr. Bras. de Geol.*, Recife, Bol. No. 1, 181.

Fairchild, T. R., Barbour, A. P. & Haralyi, N. L. E. (1978). Microfossils in the 'Eopaleozoic' Jacadigo Group at Urucum, Mato Grosso, southwest Brazil. *Bol. IG Geosciencias Univ. S. Paulo*, **9**, 74–9.

Fauchald, K. (1975). Polychaete phylogeny: a problem in protostome evolution. *Syst. Zool.*, **23**, 493–506.

Fedonkin, M. A. (1976). (Metazoan traces from the Valdai Series.) *Izv. Akad. Nauk SSSR*, Ser. Geol., No. 4, 129–32. (In Russian.)

Fedonkin, M. A. (1977). Precambrian–Cambrian ichnocoenoses of the East European Platform. *Geol. J.*, Spec. Issue No. 9, 183–94.

Fedonkin, M. A. (1978a) (Ancient trace fossils and the behavioral evolution of sediment feeders.) *Paleont. Zhurnal*, No. 2, 106–12. (In Russian.)

Fedonkin, M. A. (1978b) (A new discovery of soft-bodied Metazoa in the Vendian of the Winter Coast.) *Doklady Akad. Nauk SSSR*, **239**, 1423–6. (In Russian.)

Fedonkin, M. A. (1980a). New representatives of Precambrian coelenterates on the north of the Russian Platform. *Paleont. Zhurnal*, No. 2, 7–15. (In Russian.)

Fedonkin, M. A. (1980b). (Fossil traces of Precambrian Metazoa.) *Izv. Akad. Nauk SSSR*, Ser. Geol., No. 1, 39–46. (In Russian.)

Fedonkin, M. A. (1980c). (Early stages of evolution of Metazoa on the basis of paleoichnological data.) *Zhurnal Obshchey Biol.* **41**, 226–33. (In Russian.)

Fedonkin, M. A. (1981a). *White Sea biota of Vendian. Precambrian non-skeletal fauna of the Russian Platform North*. Trans. Geol. Inst. Acad. Sci. USSR, **342**, 100 pp. (In Russian.)

Fedonkin, M. A. (1981b). (The richest Precambrian fossil locality.) *Priroda* No. 5, 94–102. (In Russian.)

Fenchel, T. M. & Riedl, R. J. (1970). The sulfide system: a new biotic community underneath the oxidized layer of marine sand bottoms. *Marine Biol.*, **7**, 255–68.

Fenton, M. A. & Fenton, C. L. (1934). *Scolithus* as a fossil phoronid. *Pan-Am. Geol.*, **61**, 341–8.
Fischer, A. G. (1965). Fossils, early life and atmospheric history. *Proc. Nat. Acad. Sci.*, **53**, 1205–15.
Fischer, A. G. (1972). *Atmosphere and the evolution of life. Main Currents in Modern Thought*, **28**, No. 5, 9 pp.
Florkin, M. (1966). *Aspects moleculaires de l'adaptation et de la phylogénie*. Masson, Paris.
Florkin, M. & Stotz, E. H. (eds.) (1971). *Comparative biochemistry* vol. 26 pt C. Elsevier, Amsterdam.
Forbes, B. G. (1971). Stratigraphic subdivision of the Pound Quartzite (Late Precambrian, South Australia). *Trans. Roy. Soc. S. Aust.*, **95**, 219–25.
Forbes, B. G. (1972). *Explanatory notes, 1:250,000 Geological Series Sheet SH/54-9*. Geol. Survey S. Australia.
Ford, T. D. (1958). Precambrian fossils from Charnwood Forest. *Proc. Yorkshire Geol. Soc.* **31**, 211–17.
Ford, T. D. (1963). The Pre-Cambrian fossils of Charnwood Forest. *Trans. Leicester Lit. Phil. Soc.*, **57**, 57–62.
Ford, T. D. (1968). Precambrian rocks, B. The Precambrian palaeontology of Charnwood Forest. In: P. C. Sylvester-Bradley & T. D. Ford (eds.), *The geology of the east Midlands* pp. 12–14. Leicester Univ. Press, Leicester.
Ford, T. D. (1979). Precambrian fossils and the origin of the Phanerozoic phyla. In: M. R. House (ed.), *The origin of major invertebrate groups*, pp. 7–21. Academic Press, London.
Ford, T. D. (1980). The Ediacarian fossils of Charnwood Forest, Leicestershire. *Proc. Geol. Ass.*, **91**, 81–3.
Ford, T. D. & Breed, W. J. (1973). The problematical fossil *Chuaria*. *Palaeontology*, **16**, 535–50.
Foster, M. W. (1979). Soft-bodied coelenterates in the Pennsylvanian of Illinois. In: M. H. Nitecki, (ed.), *Mazon Creek fossils*, pp. 191–267. Academic Press, New York.
Fox, S. W. & Dose, K. (1972). *Molecular evolution and the origin of life*. W. H. Freeman, San Francisco.
Føyn, S. & Glaessner, M. F. (1979). *Platysolenites*, other animal fossils, and the Precambrian–Cambrian transition in Norway. *Norsk Geol. Tidsskr.*, **59**, 25–46.
Frakes, L. A. (1979). *Climates throughout geological time*. Elsevier, Amsterdam.
Francis, S., Margulis, L. & Barghoorn, E. S. (1978). On the experimental silicification of microorganisms. II. On the time of appearance of eukaryotic organisms in the fossil record. *Precambrian Res.*, **6**, 65–100.
Frey, R. W. (1975). *The study of trace fossils*. Springer, New York.
Frey, R. W. & Seilacher, A. (1980). Uniformity in marine invertebrate ichnology. *Lethaia*, **13**, 183–207.
Fritz, W. H. (1980). International Precambrian–Cambrian Working Group's field study to Mackenzie Mountains, Northwest Territories, Canada. *Current Research Part A, Geol. Survey Canada*, Pap. 80–1A, 41–5.
Fyodorov, A. B., Yegorova, L. J. & Savitsky, V. E. (1979). The first find of ancient trilobites in the lower part of the stratotype of the Tommotian Stage of the Lower Cambrian (River Aldan). *Doklady Akad. Nauk USSR*, **249**, 1188–90.
Garrett, P. (1970). Phanerozoic stromatolites: noncompetitive ecologic restriction by grazing and burrowing animals. *Science*, **169**, 171–3.
Germs, G. J. B. (1972a). *The stratigraphy and palaeontology of the Lower Nama Group, South West Africa*. Chamber of Mines Precam. Res. Unit, Dept. Geol., Univ. Cape Town, Bull. 12.

Germs, G. J. B. (1972*b*). New shelly fossils from the Nama Group, South West Africa. *Am. J. Sci.*, **272**, 752–61.

Germs, G. J. B. (1972*c*). Trace fossils from the Nama Group, South West Africa, *J. Paleont.*, **46**, 864–70.

Germs, G. J. B. (1973*a*). A reinterpretation of *Rangea schneiderhoehni* and the discovery of a related new fossil from the Nama Group, South West Africa. *Lethaia*, **6**, 1–10.

Germs, G. J. B. (1973*b*). Possible Sprigginid worms and a new trace fossil from the Nama Group, South West Africa. *Geology*, **1**, 69–70.

Glaessner, M. F. (1958). New fossils from the base of the Cambrian in South Australia. *Trans. Roy. Soc. S. Aust.*, **81**, 185–8.

Glaessner, M. F. (1959). Precambrian Coelenterata from Australia, Africa and England. *Nature*, **183**, 1472–3.

Glaessner, M. F. (1963). Zur Kenntnis der Nama-Fossilien Südwest-Afrikas. *Ann. Naturhist Mus. Wien*, **66**, 113–20.

Glaessner, M. F. (1966). Precambrian palaeontology. *Earth-Sci. Rev.*, **1**, 29–50.

Glaessner, M. F. (1969*a*). Trace fossils from the Precambrian and basal Cambrian. *Lethaia*, **2**, 369–93.

Glaessner, M. F. (1969*b*). Decapoda. In: R. C. Moore (ed.), *Treatise on invertebrate paleontology*, Pt R, Arthropoda 4, pp. R399–R533. Geol. Soc. America and Univ. Kansas.

Glaessner, M. F. (1971*a*). Geographic distribution and time range of the Ediacara Precambrian Fauna. *Geol. Soc. Am. Bull.* **82**, 509–14.

Glaessner, M. F. (1971*b*). The genus *Conomedusites* Glaessner & Wade and the diversification of the Cnidaria. *Paläont. Zeitschr.*, **43**, 7–17.

Glaessner, M. F. (1976*a*). Early Phanerozoic annelid worms and their geological and biological significance. *J. Geol. Soc. Lond.*, **132**, 259–75.

Glaessner, M. F. (1976*b*). A new genus of polychaete worms from the Late Precambrian of South Australia. *Trans. Roy. Soc. S. Aust.*, **100**, 169–70.

Glaessner, M. F. (1977). The Ediacara fauna and its place in the evolution of the Metazoa. In: A. V. Sidorenko (ed.), *Correlation of the Precambrian*, vol. 1, pp. 257–68. 'Nauka', Moscow.

Glaessner, M. F. (1978). Re-examination of *Archaeichnium*, a fossil from the Nama Group. *Ann. S. Afr. Mus.*, **74**, 335–42.

Glaessner, M. F. (1979*a*). An echiurid worm from the Late Precambrian. *Lethaia*, **12**, 121–4.

Glaessner, M. F. (1979*b*). Precambrian. In: R. A. Robison & C. Teichert (eds.), *Treatise on invertebrate paleontology*, Pt A, pp. 79–118. Geol. Soc. America and Univ. Kansas, Boulder, Colorado and Lawrence, Kansas.

Glaessner, M. F. (1980). *Parvancorina* – an arthropod from the Late Precambrian of South Australia. *Ann. Naturhist. Mus. Wien*, **83**, 83–90.

Glaessner, M. F. (1983). The emergence of Metazoa in the early history of life. *Precambrian Res.*, **20**, 427–41.

Glaessner, M. F. & Daily, B. (1959). The geology and Late Precambrian fauna of the Ediacara Fossil Reserve. *Rec. S. Aust. Mus.*, **13**, 369–401.

Glaessner, M. F. & Wade, M. (1966). The Late Precambrian fossils from Ediacara, South Australia. *Palaeontology*, **9**, 599–628.

Glaessner, M. F. & Wade, M. (1971). *Praecambridium* – a primitive arthropod. *Lethaia*, **4**, 71–7.

Glaessner, M. F. & Walter, M. R. (1975). New Precambrian fossils from the Arumbera Sandstone, Northern Territory, Australia. *Alcheringa*, **1**, 11–28.

Glaessner, M. F. & Walter, M. R. (1981). Australian Precambrian palaeobiology. In: D. R. Hunter (ed.), *Precambrian of the Southern Hemisphere*, pp. 361–96. Elsevier, Amsterdam.

Goldring, R. & Curnow, C. N. (1967). The stratigraphy and facies of the Late Precambrian of Ediacara, South Australia. *J. Geol. Soc. Aust.*, **14**, 195–214.

Gould, S. J. (1970). Evolutionary paleontology and the science of form. *Earth-Sci. Rev.*, **6**, 77–119.

Gould, S. J. (1977). *Ontogeny and phylogeny.* Harvard Univ. Press, Cambridge, Mass.

Gould, S. J., Raup, D. M., Sepkoski, J. J., Schopf, T. J. M. & Simberloff, D. S. (1977). The shape of evolution: a comparison of real and random clades. *Paleobiology*, **3**, 23–40.

Grasshoff, M. (1981). Arthropodisierung als biomechanischer Prozess und die Entstehung der Trilobiten-Konstruktion. *Paläont. Zeitschr.* **55**, 219–35.

Grell, K. G. (1973). *Protozoology.* Springer Verlag, New York.

Gutmann, W. F. (1965). Rückstoss und Ruderschlag der Quallen. *Natur u. Museum*, **95**, 455–63.

Gutmann, W. F. (1977). Phylogenetic reconstruction: theory, methodology and application to chordate evolution. In: M. K. Hecht, P. C. Goody & B. M. Hecht (eds.), *Major patterns in vertebrate evolution*, pp. 645–69. Plenum Press, New York.

Gutmann, W. F. (1981). Relationship between invertebrate phyla based on functional–mechanical analysis of the hydrostatic skeleton. *Am. Zool.*, **21**, 63–81.

Gutmann, W. F., Vogel, K. & Zorn, W. (1978). Brachiopods: biomechanical interdependences governing their origin and phylogeny. *Science*, **199**, 890–3.

Hadži, J. (1963). *The evolution of the Metazoa.* Pergamon Press, Oxford.

Hahn, G. & Pflug, H. D. (1980). Ein neuer Medusen-Fund aus dem Jung-Präkambrium von Zentral-Iran. *Senckenbergiana Lethaea*, **60**, 449–61.

Hakes, W. G. (1976). Trace fossils and depositional environment of four clastic units, Upper Pennsylvanian megacylothems, northeast Kansas. *Univ. Kansas Paleont. Contrib.*, Art. 63.

Haldane, J. B. S. (1929). The origin of life. *Rationalist Annual* (reprinted in Bernal (1967), pp. 242–9).

Hallam, A. (ed.) (1977). *Patterns of evolution, as illustrated by the fossil record. Developments in palaeontology and stratigraphy*, vol. 5. Elsevier, Amsterdam.

Hambrey, M. J. & Harland, W. B. (1981). *Earth's pre-Pleistocene glacial record.* Cambridge Univ. Press, Cambridge, England.

Hand, C. (1959). On the origin and phylogeny of the coelenterates. *Syst. Zool.*, **8**, 191–202.

Hanson, E. D. (1977). *The origin and early evolution of animals.* Wesley Univ. Press, Middleton, Conn.

Häntzschel, W. (1975) Trace fossils and problematica. In: C. Teichert (ed.), *Treatise on invertebrate paleontology*, pt W, suppl. 1, 269 pp. Geol. Soc. America and Univ. Kansas.

Harland, W. B. (1978). Geochronologic scales. *Contributions to the geologic time scale. Am. Ass. Petroleum Geol., Studies in Geology*, No. 6, 9–32.

Harland, W. B. & Herod, K. N. (1975). Glaciations through time. *Geol. J.* Spec. Issue 6, 189–216.

Harrison, J. E. & Peterman, Z. E. (1982). North American Commission on Stratigraphic Nomenclature. Report 9, Adoption of geochronometric units for division of Precambrian time. *Am. Ass. Petroleum Geol. Bull.* **66**, 801–4.

Heberer, G. (1967–1971). *Die Evolution der Organismen* (3rd edn), 3 vols. Gustav Fischer Verlag, Stuttgart.

Hedberg, H. D. (1974). Basis for chronostratigraphic classification of the Precambrian. *Precambrian Res.* **1**, 165–77.

Henderson, L. J. (1913). *The fitness of the environment.* Macmillan. (Reprinted Beacon Press, Boston, 1958.)

Hofmann, H. J. (1971). *Precambrian fossils, pseudofossils and problematica in Canada. Geol. Survey Canada, Bull.* **189**, 146 pp.

Hofmann, H. J. (1972). Precambrian remains in Canada: fossils, dubiofossils and pseudofossils. *Int. Geol. Congr., 24th Sess., Sect. 1*, 20–30. Montreal.

Hofmann, H. J. (1977). The problematic fossil *Chuaria* from the Late Precambrian Uinta Mountain Group, Utah. *Precambrian Res.*, **4**, 1–11.

Hofmann, H. J. (1981). First find of a Late Proterozoic faunal assemblage in the North American Cordillera. *Lethaia*, **14**, 303–10.

Hofmann, H. J. & Aitken, J. D. (1979). Precambrian biota from the Little Dal Group, Mackenzie Mountains, northwestern Canada. *Canadian J. Earth-Sci.*, **16**, 150–66.

Hofmann, H. J., Hill, J. & King, A. F. (1979). Late Precambrian microfossils, southeastern Newfoundland. In: *Current Research, Part B, Geol. Survey Canada*, Paper 79–1B, 83–98.

Holland, H. D. (1978). *The chemistry of the atmosphere and oceans.* Wiley-Interscience, New York.

House, M. R. (ed.) (1979). *The origin of major invertebrate groups* (Syst. Ass. Spec. vol. No. 12). Academic Press, London.

Houzay, J.-P. (1979). Empreintes attribuables à des méduses dans la série de base de l'Adoudounien (Précambrien terminal de l'Anti-Atlas, Maroc). *Géol. Médit.*, **6**, 379–84.

Hyman, L. H. (1940). *The invertebrates*, vol. 1, *Protozoa through Ctenophora.* McGraw Hill, New York.

Hyman, L. H. (1959). *The invertebrates*, vol. 5, *Smaller coelomate groups.* McGraw Hill, New York.

International Stratigraphic Guide (1976). Int. Subcomm. Strat. Classif., (H. D. Hedberg, ed.). Wiley-Interscience, New York.

Ivanov, A. V. (1968). (*The origin of multicellular animals.*) 'Nauka', Leningrad. (In Russian.)

Ivanov, A. V. (1970). Verwandtschaft und Evolution der Pogonophoren. *Z. Zool. Syst. Evolut.-forsch.*, **8**, 109–19.

Jaeger, H. & Martinsson, A. (1980). The Early Cambrian trace fossil *Plagiogmus* in its type area. *Geol. Fören. Stockholm Förh.*, **102**, 117–26.

Jägersten, G. (1972). *Evolution of the metazoan life cycle.* Academic Press, London.

James, H. L. (1978) Subdivision of the Precambrian – a brief review and a report on recent decisions by the Subcommission on Precambrian Stratigraphy. *Precambrian Res.*, **7**, 193–204.

Jell, P. A. (1979). *Plumulites* and the machaeridian problem. *Alcheringa*, **3**, 253–9.

Jell, P. A. (1981). *Thambetolepis delicata* gen. sp. nov., an enigmatic fossil from the Early Cambrian of South Australia. *Alcheringa*, **5**, 85–93.

Jenkins, R. J. F. (1981). The concept of an Ediacarian Period and its stratigraphic significance in Australia. *Trans. Roy. Soc. S. Aust.*, **105**, 179–94.

Jenkins, R. J. F., Ford, C. H. & Gehling, J. G. (1983). The Ediacara Member of the Rawnsley Quartzite: the context of the Ediacara assemblage (Late Precambrian, Flinders Ranges). *J. Geol. Soc. Aust.*, **29**, 101–119.

Jenkins, R. J. F. & Gehling, J. G. (1978). A review of the frond-like fossils of the Ediacara assemblage. *Rec. S. Aust. Mus.*, **17**, 347–59.

Jenkins, R. J. F., Plummer, P. S. & Moriarty, K. C. (1981). Late Precambrian pseudofossils from the Flinders Ranges, South Australia. *Trans. Roy. Soc. S. Aust.*, **105**, 67–83.

Kauffman, E. G. & Steidtmann, R. (1981). Are these the oldest metazoan trace fossils? *J. Paleont.*, **55**, 923–47.

Keller, B. M. (1979). Precambrian stratigraphic scale of the U.S.S.R. *Geol. Mag.*, **116**, 419–29.

Keller, B. M. & Fedonkin, M. A. (1976). (New finds of fossils in the Valdai Series, Precambrian, R. Syuzma.) *Izv. Akad. Nauk SSSR, Ser. Geol.*, No. 3, 38–44. (In Russian.)

Kent, L. E. & Hugo, P. J. (1978). Aspects of the revised South African stratigraphic classification and a proposal for the chronostratigraphic subdivision of the Precambrian. *Contributions to the geologic time scale. Am. Ass. Petroleum Geol., Studies in Geology*, No. 6, 367–79.

Knoll, A. H. & Barghoorn, E. S. (1975). Precambrian eucaryote organisms: a reassessment of the evidence. *Science*, **190**, 52–4.

Knoll, A. H. & Golubic, S. (1979). Anatomy and taphonomy of a Precambrian algal stromatolite. *Precambrian Res.*, **10**, 115–51.

Knoll, A. H. & Vidal, G. (1980). Late Proterozoic vase-shaped microfossils from the Visingsö Beds, Sweden. *Geol. Fören. Stockholm Förh.*, **102**, 207–11.

Kowalski, W. R. (1978). Critical analysis of Cambrian ichnogenus *Plagiogmus* Roedel. *Ann. Soc. Géol. Pologne*, **48**, 333–44.

Kröner, A. (1977). Non-synchroneity of Late Precambrian glaciations in Africa. *J. Geol.*, **85**, 289–300.

Kröner, A. & Rankama, K. (1972). *Late Precambrian glaciogenic sedimentary rocks in southern Africa: a compilation with definitions and correlations.* Chamber of Mines Precambrian Res. Unit, Dept. Geol., Univ. Cape Town, Bull. 11, 37 pp.

Kröner, A., McWilliams, M. O., Germs, G. J. B., Reid, A. B. & Schalk, K. E. L. (1980). Palaeomagnetism of Late Precambrian to Early Paleozoic mixtite-bearing formations in Namibia (South-west Africa): the Nama Group and Blaubeker Formation. *Am. J. Sci.*, **280**, 942–68.

Kulling, O. (1972). The Swedish Caledonides. In: T. Strand & O. Kulling, (eds.), *Scandinavian Caledonides*, pp. 147–302. Wiley-Interscience, London.

Larwood, G. P. & Taylor, P. D. (1979). Early structural and ecological diversification in the Bryozoa. In: M. R. House (ed.), *The origin of major invertebrate groups*, pp. 209–34. Academic Press, London.

Lauterbach, K. E. (1973). Schlüsselereignisse in der Evolution der Stammgruppe der Euarthropoda. *Zool. Beitr.*, N. F. **19**, 251–99.

Lauterbach, K. E. (1974). Uber die Herkunft des Carapax der Crustaceen. *Zool. Beitr.* N.F. **20**, F273–327.

Lauterbach, K. E. (1978). Gedanken zur Evolution der Euarthropoden-Extremität. *Zool. Anat. Jb.*, **99**, 64–92.

Lauterbach, K. E. (1980*a*). Schlüsselereignisse in der Evolution des Grundplans der Mandibulata (Arthropoda). *Abh. Naturwiss. Ver. Hamburg*, N. F., **23**, 105–61.

Lauterbach, K. E. (1980*b*). Schlüsselereignisse in der Evolution des Grundplans der Arachnata (Arthropoda). *Abh. naturwiss. Ver. Hamburg*, N. F. **23**, 163–327.

Leeson, B. (1970). Geology of the Beltana 1:63,360 map area. *Rep. Invest. Geol. Survey S. Aust.*, **27**, 19 pp.

Liu Xiaoliang (1981). Metazoan fossils from the Mashan Group near Jixi, Heilongjiang. *Bull. Chinese Acad. Geol. Sci.*, **3**, 71–83. (In Chinese.)

Lovelock, J. E. (1979). *Gaia: a new look at life on earth.* Oxford Univ. Press, Oxford.

Lovelock, J. E. & Margulis, L. (1974). Atmospheric homeostasis, by and for the biosphere: the Gaia hypothesis. *Tellus*, **26**, 1–10.

Lowenstam, H. A. (1980). What, if anything, happened in the transition from the Precambrian to the Phanerozoic? *Precambrian Res.*, **11**, 89–91.

Lowenstam, H. A. & Margulis, L. (1980). Evolutionary prerequisites for Early Phanerozoic calcareous skeletons. *BioSystems*, **12**, 27–41.

McKirdy, D. M. (1974). Organic geochemistry in Precambrian research. *Precambrian Res.*, **1**, 75–137.

McLaughlin, P. J. & Dayhoff, M. O. (1973). Eukaryotic evolution: a view based on cytochrome c sequence data. *J. Molec. Evol.* **2**, 99–116.
McWilliams, M. O. & McElhinny, M. W. (1980). Late Precambrian paleomagnetism of Australia: the Adelaide Geosyncline. *J. Geol.*, **88**, 1–26.
Major, R. B. (1974). The Punkerri Beds. *Quart. Geol. Notes, Geol. Survey S. Aust.*, No. 51, 2–5.
Manton, S. M. (1967). The polychaete *Spinther* and the origin of the Arthropoda. *J. nat. Hist.*, **1**, 1–22.
Marek, L. & Yochelson, E. L. (1976). Aspects of the biology of Hyolitha (Mollusca). *Lethaia*, **9**, 65–82.
Margulis, L. (1970). *Origin of eukaryotic cells*. Yale Univ. Press, New Haven.
Margulis, L. (1974). Classification and evolution of prokaryotes and eukaryotes. In: R. King (ed.), *Handbook of genetics*, vol. 1, pp. 1–41. Plenum Press, New York.
Margulis, L. (1975). Symbiotic theory of the origin of eukaryotic organelles. Criteria for proof. In: *Symp. Soc. Exp. Biol.*, vol. 29, *Symbiosis*, pp. 21–38. Cambridge University Press, Cambridge, England.
Margulis, L. & Lovelock, J. E. (1974). Biological modulation of the earth's atmosphere. *Icarus*, **21**, 471–89.
Margulis, L. & Lovelock, J. E. (1978). The biota as ancient and modern modulator of the earth's atmosphere. *Pure & Appl. Geophys.*, **116**, 239–43.
Margulis, L., Walker, J. C. G. & Rambler, M. (1976). Re-assessment of the roles of oxygen and ultraviolet light in Precambrian evolution. *Nature*, **264**, 620–4.
Martin, H. (1965). *The Precambrian geology of South West Africa and Namaqualand*. Precambrian Res. Unit. Univ. Cape Town.
Matthews, S. C. & Missarzhevsky, V. V. (1975). Small shelly fossils of Late Precambrian and Early Cambrian age: a review of recent work. *J. Geol. Soc. Lond.*, **131**, 289–304.
Mawson, D. (1938). Cambrian and sub-Cambrian formations at Parachilna Gorge. *Trans. Roy. Soc. S. Aust.*, **26**, 255–62.
Mayr, E. (1963). *Animal species and evolution*. Harvard Univ. Press, Cambridge, Mass.
Mayr, E. (1982). *The growth of biological thought: diversity, evolution and inheritance*. Harvard Univ. Press, Cambridge, Mass.
Meneisy, M. Y. & Miller, J. A. (1963). A geochemical study of the crystalline rocks of Charnwood Forest, England. *Geol. Mag.*, **100**, 507–23.
Mettam, L. (1971). Functional design and evolution of the polychaete *Aphrodite aculeata*. *J. Zool., Lond.* **163**, 489–514.
Miller, S. L. (1953). A production of aminoacids under possible primitive earth conditions. *Science*, **117**, 528–9.
Miller, S. L. & Urey, H. C. (1959). Organic compound synthesis on the primitive earth. *Science*, **130**, 245–51.
Mincham, H. (1958). The oldest fossils found in South Australia. *Education Gazette, S. Aust.*, **74**, 216–19.
Misra, S. B. (1969). Late Precambrian (?) fossils from southeastern Newfoundland. *Geol. Soc. Am. Bull.*, **80**, 2133–40.
Missarzhevsky, V. V. (1973). (Conodontomorphs of the boundary beds of Cambrian and Precambrian of Siberian Platform and Kazakhstan.) In: I. T. Zhuravleva (ed.), *Problems of the Lower Cambrian palaeontology and biostratigraphy of Siberia and Far East*. pp. 53–7. 'Nauka', Novosibirsk. (In Russian.)
Missarzhevsky, V. V. (1980). (Boundary layers of the Cambrian and Precambrian, western slope of Olenek Highlands.) *Bull. Soc. Naturalistes Moscow*, **55**, 23–34. (In Russian.)
Missarzhevsky, V. V. & Rozanov, A. Yu. (1968). Tommotian Stage and the problem of the Lower Paleozoic boundary. *Int. Geol. Congr., 23rd Sess.*, 40–50. (Doklady Sov. Geol. Probl. 9, 'Nauka', Moscow.) (In Russian.)

Monod, J. (1970). *Le hasard et la necessité*. Editions du Seuil, Paris.
Morel, P. & Irving, E. (1978). Tentative palaeocontinental maps for the Early Phanerozoic and Proterozoic. *J. Geol.*, **86**, 535–61.
Müller, A. H. (1979). Fossilization (Taphonomy) In: R. A. Robison & C. Teichert (eds.), *Treatise on invertebrate paleontology*, Pt A, pp. 1–78. Geol. Soc. America and Univ. Kansas, Boulder, Colorado, and Lawrence, Kansas.
Müller, K. J. (1979). Phosphatocopine ostracodes with phosphatized appendages from the Upper Cambrian of Sweden. *Lethaia*, **12**, 1–27.
Müller, K. J. (1981). Arthropods with phosphatized soft parts from the Upper Cambrian 'orsten' of Sweden. *Short Pap. 2nd Int. Symp. Cambrian Syst.* Open-File Rep. 81–743, 147–151. (United States Geological Survey.)
Nitecki, M. G., Zhuravleva, I. T., Myagkova, Ye. I. & Toomey, D. F. (1981). Similarity of *Soanites bimuralis* to Archaeocyatha and receptaculitids. *Paleont. Zhurnal*, No. 1, 5–9. (In Russian. For English transl., see *Paleont J.*, **15**, 1–5.)
Nursall, J. R. (1962). On the origins of the major groups of animals. *Evolution*, **16**, 118–23.
Oehler, J. H. (1976). Experimental studies in Precambrian paleontology: structural and chemical changes in blue-green algae during simulated fossilization in synthetic chert. *Geol. Soc. Am. Bull.*, **87**, 117–29.
Oehler, J. H., Oehler, D. Z. & Muir, M. D. (1976). On the significance of tetrahedral tetrads of Precambrian algal cells. *Origins of Life*, **7**, 259–67.
Oparin, A. I. (1924). (*The origin of life*.) Moskovsky Rabotchy, Moscow. (In Russian.)
Öpik, A. A. (1975). *Cymbric Vale fauna of New South Wales and Early Cambrian biostratigraphy*. Bureau Min. Res., Geol. Geophys. Bull., 159.
Palij, V. M. (1974). On finding of the trace fossil in the Riphean deposits of the Ovrutch Ridge. *Rep. Acad. Sci. Ukr. SSR*, Ser. B., Geol. Geophys., etc., **36**, 34–7. (In Ukrainian.)
Palij, V. M., Posti, E. & Fedonkin, M. A. (1979). Soft-bodied Metazoa and trace fossils of Vendian and Lower Cambrian. *Upper Precambrian and Cambrian paleontology of East European Platform*, Acad. Sci. USSR Soviet–Polish Working Group, 49–82. (In Russian.)
Pantin, C. F. A. (1951). Organic design. *Advancement Sci.*, **8**, 138–50.
Patchett, P. J., Gale, N. H., Goodwin, R. & Humm, M. J. (1980). Rb–Sr whole rock isochron ages of Late Precambrian to Cambrian igneous rocks from southern Britain. *J. Geol. Soc. Lond.*, **137**, 649–56.
Patterson, C. (1978). *Evolution*. British Mus. Nat. Hist., London.
Paul, C. R. (1977). Evolution of primitive echinoderms. In: A. Hallam (ed.), *Patterns of evolution, as illustrated by the fossil record*, pp. 123–58. Elsevier, Amsterdam.
Peters, D. S. (1972). Das Problem konvergent entstandener Strukturen in der anagenetischen und genealogischen Systematik. *Z. Zool. Syst. Evol.-Forsch.*, **10**, 161–73.
Peters, D. S. & Gutmann, W. F. (1971). Uber die Lesrichtung von Merkmals- und Konstruktionsreihen. *Z. Zool. Syst. Evol.-Forsch.*, **9**, 237–63.
Pflug, H. D. (1966) Einige Reste niederer Pflanzen aus dem Algonkium. *Palaeontographica*, Abt. *B*, **117**, 59–74.
Pflug, H. D. (1970*a*). Zur Fauna der Nama-Schichten in Südwest-Afrika. I. Pteridinia, Bau und systematische Zugehörigkeit. *Palaeontographica*, Abt. *A*, **134**, 153–262.
Pflug, H. D. (1970*b*). Zur Fauna der Nama-Schichten in Südwest-Afrika. II. Rangeidae, Bau und systematische Zugehörigkeit. *Palaeontographica*, Abt. *A*, **135**, 198–231.
Pflug, H. D. (1972). Zur Fauna der Nama-Schichten in Südwest-Afrika. III. Erniettomorpha, Bau und Systematik. *Palaeontographica*, Abt. *A*, **139**, 134–70.
Pflug, H. D. (1973). Zur Fauna der Nama-Schichten in Südwest-Afrika. IV. Mikroskopische Anatomie der Petalo-Organismen. *Palaeontographica*, Abt. *A*, **144**, 166–202.

Pojeta, J. (1980). Molluscan phylogeny. *Tulane Studies in Geol. and Pal.* **16**, 55–80.
Poulsen, V. (1963). *Notes on* Hyolithellus *Billings*, 1871, *Class Pogonophora Johansson, 1937. Biol. Medd. Danske Vid. Selsk.*, **23**, No. 12, 15 pp.
Pringle, I. R. (1973). Rb–Sr age determinations on shales associated with Varanger ice age. *Geol. Mag.*, **199**, 465–72.
Raff, R. A. & Raff, E. C. (1970). Respiratory mechanisms and the metazoan fossil record. *Nature*, **228**, 1003–5.
Ratner, M. I. & Walker, J. C. G. (1972). Atmospheric ozone and the history of life. *J. Atmospheric Sci.*, **29**, 803–8.
Reisinger, E. (1970). Zur Problematik der Evolution der Coelomaten. *Z. Zool. Syst. Evol.-Forsch.*, **8**, 81–109.
Remane, A. (1956). *Die Grundlagen des natürlichen Systems, der vergleichenden Anatomie und der phylogenetischen Morphologie und Systematik* (2nd edn.). Akad. Verlagsges, Leipzig.
Remane, A. (1973). Stellungnahme. In: *Das Archicoelomatenproblem, Aufsätze u. Reden Senckenb. Naturf. Ges.*, No. 22, pp. 105–8. Frankfurt.
Repetski, J. E. & Szaniawski, H. (1981). Paleobiologic interpretation of Cambrian and earliest Ordovician conodont natural assemblages. *Short Pap. 2nd Int. Symp. Cambrian Syst.*, Open-File Rep. 81–743, 169–72. (United States Geological Survey.)
Rhoads, D. C. (1974). Organism–sediment relations in the muddy sea floor. *Oceanogr. Mar. Biol. Ann. Rev.*, **12**, 263–300.
Rhoads, D. C. & Morse, J. W. (1971). Evolutionary and ecologic significance of oxygen-deficient marine basins. *Lethaia*, **4**, 413–28.
Richter, R. (1955). Die ältesten Fossilien Süd-Afrikas. *Senckenbergiana Lethaea*, **36**, 243–89.
Riedl, L. (1978). *Order in living organisms* (*A systems analysis of evolution*). J. Wiley & Sons, Chichester, New York.
Rieger, R. M. & Sterrer, W. (1975). New spicular skeletons in Turbellaria and the occurrence of spicules in marine meiofauna. Part 1. *Sonderh. Z. Zool. Syst. Evol.-Forsch.*, **13**, 207–48.
Rieppel, O. (1980). Homology, a deductive concept? *Z. Zool. Syst. Evol.-Forsch.*, **18**, 315–19.
Roberts, J. D. (1976). Late Precambrian dolomites, Vendian glaciation, and synchroneity of Vendian glaciations. *J. Geol.*, **84**, 47–63.
Rowell, A. J. (1981). The Cambrian brachiopod radiation – monophyletic or polyphyletic origins? *Short Pap. 2nd Int. Symp. Cambrian Syst.*, Open-File Rep. 81–743, 184–7. (United States Geological Survey.)
Rozanov, A. Yu. (1973). *Regularities in the morphological evolution of regular archaeocyathean and the problem of the Lower Cambrian Stage division. Acad. Sci. USSR, Geol. Inst. Trans.*, **241**, 164 pp. (In Russian.)
Rozanov, A. Yu. (1976a). Biogeography and stages of the Early Cambrian. *Acad. Sci. USSR, Int. Geol. Congr, 25th Sess., Reports of Soviet Geologists. Palaeont., Marine Geol.* 'Nauka', Moscow. (In Russian.)
Rozanov, A. Yu. (1976b). Precambrian–Cambrian boundary. In: A. N. Peive (ed.), *Boundaries of geological systems*, pp. 31–53. 'Nauka', Moscow. (In Russian.)
Rozanov, A. Yu. & Missarzhevsky, V. V. (1966). *Biostratigraphy and fauna of Lower Cambrian Horizons. Acad. Sci. USSR, Geol. Inst. Trans.*, **148**, 126 pp (In Russian.)
Rozanov, A. Yu., Missarzhevsky, V. V., Volkova, N. A., Voronova, L. G., Krylov, I. N., Keller, B. M., Korolyuk, I. K., Lendzion, K., Michniak, R., Pychova, N. G. & Sidorov, A. D. (1969). *Tommotian Stage and the Cambrian Lower boundary problem. Acad. Sci. USSR, Geol. Inst. Trans.*, **206**, 380 pp. (In Russian.)
Runnegar, B. (1980a). Hyolitha: status of the phylum. *Lethaia*, **13**, 21–5.

Runnegar, B. (1980b). Mollusca: the first hundred million years. *J. Malac. Soc. Aust.*, **4**, 223–4. (Abstract.)

Runnegar, B. (1982a). Oxygen requirements, biology and phylogenetic significance of the Late Precambrian worm *Dickinsonia*, and the evolution of the burrowing habit. *Alcheringa*, **6**, 223–39.

Runnegar, B. (1982b). The Cambrian explosion: animals or fossils? *J. Geol. Soc. Aust.*, **29**, 395–411.

Runnegar, B. & Jell, P. A. (1976). Australian Middle Cambrian molluscs and their bearing on early molluscan evolution. *Alcheringa*, **1**, 109–38.

Runnegar, B. & Pojeta, J. (1974). Molluscan phylogeny: the paleontological viewpoint. *Science*, **186**, 311–17.

Runnegar, B., Pojeta, J., Taylor, J. D., Taylor, M. E. & McClung, G. (1975). Biology of the Hyolitha. *Lethaia*, **8**, 181–91.

Rutten, M. G. (1962). *The geological aspects of the origin of life on earth.* Elsevier, Amsterdam.

Rutten, M. R. (1966). Geologic data on atmospheric history. *Palaeogeogr., Palaeoclimat., Palaeoecol.*, **2**, 47–57.

Sabrodin, W. (1971). Leben im Präkambrium. *Ideen des exakten Wissens*, 12/71, 835–42.

Sabrodin, W. (1972). Leben im Präkambrium. *Bild der Wissenschaft*, 1972, 586–91.

Salvini-Plawen, L. v. (1969). Solenogastres und Caudofoveata (Mollusca, Aculifera): Organisation und phylogenetische Bedeutung. *Malacologia*, **9**, 191–216.

Salvini-Plawen, L. v. (1978). On the origin and evolution of the lower Metazoa. *Z. Zool. Syst. Evol.-forsch.*, **16**, 40–88.

Scheltema, A. H. (1978). Position of the class Aplacophora in the phylum Mollusca. *Malacologia*, **17**, 99–109.

Schopf, J. W. (1975a). Precambrian paleobiology: problems and perspectives. *Ann. Rev. Earth Planet. Sci.*, **3**, 213–49.

Schopf, J. W. (1975b). The age of microscopic life. *Endeavour*, **34**, 51–8.

Schopf, J. W. (1977). Biostratigraphic usefulness of stromatolitic Precambrian microbiotas: a preliminary analysis. *Precambrian Res.*, **5**, 143–73.

Schopf, J. W. (1978). The evolution of the earliest cells. *Sci. Am.* **239**, 110–38.

Schopf, J. W. & Barghoorn, E. S. (1967). Alga-like fossils from the Early Precambrian of South Africa. *Science*, **156**, 508–12.

Schopf, J. W., Dolnik, T. A., Krylov, I. N., Mendelson, C. V., Nazarov, B. B., Nyberg, A. V., Sovietov, Yu. K. & Yakshin, M. S. (1977). Six new stromatolitic microbiotas from the Proterozoic of the Soviet Union. *Precambrian Res.*, **4**, 269–84.

Schopf, J. W., Haugh, B. N., Molnar, R. E. & Satterthwait, D. F. (1973). On the development of metaphytes and metazoans. *J. Paleont.*, **47**, 1–9.

Schopf, J. W. & Oehler, D. Z. (1976). How old are the eukaryotes? *Science*, **193**, 47–9.

Schopf, J. W. & Prasad, K. N. (1978). Microfossils in Collenia-like stromatolites from the Proterozoic Vempalle Formation of the Cuddapah Basin, India. *Precambrian Res.*, **6**, 347–66.

Schopf, T. J. M. (1980). *Paleooceanography.* Harvard Univ. Press, Cambridge, Mass.

Scrutton, C. T. (1979). Early fossil cnidarians. In: M. R. House (ed.), *The origin of major invertebrate groups*, pp. 161–207. Academic Press, London.

Seilacher, A. (1977). Evolution of trace fossil communities. In: A. Hallam (ed.), *Patterns of evolution, as illustrated by the fossil record*, pp. 357–76. Elsevier, Amsterdam.

Sepkoski, J. J. (1978). A kinetic model of Phanerozoic taxonomic diversity. I. Analysis of marine orders. *Paleobiology*, **4**, 223–51.

Sepkoski, J. J. (1979). A kinetic model of Phanerozoic taxonomic diversity. II. Early Phanerozoic families and multiple equilibria. *Paleobiology*, **5**, 222–51.

Sewertzoff, A. N. (1931). *Morphologische Gesetzmässigkeiten der Evolution.* G. Fischer Verlag, Jena.

Siewing, R. (1972). Zur Descendenz der Chordaten. Erwiderung und Versuch einer Geschichte der Archicoelomaten. *Z. Zool. Syst. Evol.-Forsch.*, **10**, 267–91.
Simpson, G. G. (1944). *Tempo and mode in evolution*. Columbia Univ. Press, New York.
Simpson, G. G. (1949). *The meaning of evolution*. Yale Univ. Press, New Haven.
Simpson, G. G. (1953). *The major features of evolution*. Columbia Univ. Press, New York.
Simpson, G. G. (1961). *Principles of animal taxonomy*. Columbia Univ. Press, New York.
Sims, P. K. (1980). Subdivision of the Proterozoic and Archean Eons: recommendations and suggestions by the International Subcommission on Precambrian Stratigraphy. *Precambrian Res.*, **13**, 379–80.
Sokolov, B. S. (1952). (On the age of the oldest sedimentary cover of the Russian Platform.) *Izv. Akad. Nauk SSSR, Ser. Geol.*, No. 5, 21–31. (In Russian.)
Sokolov, B. S. (1958). (The problem of the lower boundary of the Palaeozoic and the oldest deposits of the pre-Sinian platform of Eurasia.) *Trudy VNIGRI*, No. 126, 5–67. (In Russian; a shorter version was published in *Colloques Int. Centre Nat. Rech. Sci., Paris*, **76**, 103–28.)
Sokolov, B. S. (1967). (The oldest Pogonophora.) *Doklady Akad. Nauk SSSR*, **177**, 252–5. (In Russian.)
Sokolov, B. S. (1968). (Vendian and Early Cambrian Sabelliditida (Pogonophora) of the USSR.) *Int. Geol. Congr.*, 23rd Sess., 73–8. (Doklady Sov. Geol., Probl. 9, 'Nauka', Moscow.) (In Russian.)
Sokolov, B. S. (1972a). Vendian and Early Cambrian Sabelliditida (Pogonophora) of the USSR, *Proc. Int. Paleont. Union, 23rd Int. Geol. Congr.* (1968), pp. 79–86. Instytut Geologiczny, Warsaw.
Sokolov, B. S. (1972b). Vendian Stage in the Earth History. *Int. Geol. Congr.*, 24th Sess., 114–23. (Doklady Sov. Geol., Probl. 7, 'Nauka', Moscow.) (In Russian.)
Sokolov, B. S. (1972c). The Vendian Stage in Earth History. *Int. Geol. Congress*, 24th Sess., Section 1, 78–84. Montreal.
Sokolov, B. S. (1972d). (The Precambrian biosphere in the light of palaeontological data.) *Vestnik Akad. Nauk SSSR*, No. 8, 48–54. (In Russian.)
Sokolov, B. S. (1973). Vendian of northern Eurasia. *Am. Ass. Petroleum Geol., Arctic Geology*, Mem. 19, 204–18.
Sokolov, B. S. (1974). (The problem of the Precambrian–Cambrian boundary.) *Geol. Geophys., Sib. Otd. Akad. Nauk SSSR*, No. 2, 3–29. (In Russian.)
Sokolov, B. S. (1976a). (Precambrian Metazoa and the Vendian–Cambrian boundary). *Paleont. Zhurnal*, No. 1, 13–18. (In Russian.)
Sokolov, B. S. (1976b). (The organic world of the earth on the way to Phanerozoic differentiation.) *Vestnik Akad. Nauk SSSR*, No. 1, 126–43. (In Russian.)
Sokolov, B. S. (1977). (Perspectives of Precambrian biostratigraphy). *Akad. Nauk SSSR, Sib. Otd., Geol. Geofi*, No. 11, 54–70. (In Russian.)
Sokolov, B. S. (1978). (Some questions of the stratigraphy of the Upper Precambrian and the position of the Vendian.) In: *Problems of stratigraphy and historical geology*, pp. 20–9. Moscow University. (In Russian.)
Sokolov, B. S. & Khomentovsky, V. V. (eds.) (1975). *Analogues of the Vendian Complex in Siberia*. 'Nauka', Moscow. (In Russian.)
Sprigg, R. C. (1947). Early Cambrian (?) jellyfishes from the Flinders Ranges, South Australia. *Trans. Roy. Soc. S. Aust.*, **71**, 212–24.
Sprigg, R. C. (1949). Early Cambrian 'Jellyfishes' of Ediacara, South Australia, and Mount John, Kimberley District, Western Australia. *Trans. Roy. Soc. S. Aust.*, **73**, 72–99.
Squire, A. D. (1973). Discovery of Late Precambrian trace fossils in Jersey, Channel Islands. *Geol. Mag.*, **110**, 223–6.
Stanier, R., Adelberg, E. & Doudoroff, M. (1963). *The microbial world*. Prentice-Hall, Englewood Cliffs, N.J.

Stanley, S. M. (1973). An ecological theory for the sudden origin of multicellular life in the Late Precambrian. *Proc. Nat. Acad. Sci.*, **70**, 1486–9.

Stanley, S. M. (1976a). Fossil data and the Precambrian–Cambrian evolutionary transition. *Am. J. Sci.*, **276**, 56–76.

Stanley, S. M. (1976b). Ideas on the timing of metazoan diversification. *Paleobiology*, **2**, 209–19.

Stanley, S. M. (1979). *Macroevolution. Pattern and process.* W. H. Freeman & Co., San Francisco.

Stasek, C. R. (1972). The molluscan framework. In: M. Florkin & B. T. Scheer (eds.), *Chemical zoology*, vol. 7, pp. 1–44. Academic Press, New York.

Stöcklin, J. (1968). Structural history and tectonics of Iran: a review. *Am. Ass. Petroleum Geol. Bull.*, **52**, 1229–58.

Stöcklin, J. (1972). Iran. *Lexique Stratigraphique International*, vol. III, 9b, 1.

Szaniawski, H. (1982). Chaetognath grasping spines recognized among Cambrian protoconodonts. *J. Paleont.*, **56**, 806–10.

Tankard, A. J., Jackson, M. P. A., Eriksson, K. A., Hobday, D. K., Hunter, D. R. & Minter, W. E. L. (1982). *Crustal evolution of southern Africa.* Springer Verlag, New York.

Tappan, H. N. (1980). *The paleobiology of plant protists.* W. H. Freeman & Co., San Francisco.

Tappan, H. & Loeblich, A. R. (1971). Geobiologic implications of fossil phytoplankton evolution and time–space distribution. In: R. Kosanke & A. T. Cross (eds.), *Symposium on palynology of the Late Cretaceous and Early Tertiary. Spec. Pap. Geol. Soc. Am.*, No. 127, 247–340.

Tarling, D. H. (1978). *Evolution of the earth's crust.* Academic Press, London.

Taylor, F. J. R. (1974). Implications and extensions of the serial endosymbiosis theory of the origin of the eukaryotes. *Taxon*, **23**, 5–34.

Termier, H. & Termier, G. (1960). L'Ediacarien, premier étage paléontologique. *Rev. Gén. Sci.*, **67**, 79–87.

Thiel, M. E. (1978). Die postephyrale Entwicklung des Gastrovascularsystems der Rhizostomida nebst Ergänzungen und Berichtigungen zu den Stiasnyschen Typen dieser Entwicklung, etc. *Z. Zool. Syst. Evolut.-Forsch.*, **16**, 267–89.

Towe, K. M. (1970). Oxygen–collagen priority and the early metazoan fossil record. *Proc. Nat. Acad. Sci.*, **65**, 781–8.

Towe, K. M. (1976). Fossil data and the Precambrian–Cambrian evolutionary transition. *Am. J. Sci.*, **276**, 1178–80.

Towe, K. M. (1981). Biochemical keys to the emergence of complex life. In: J. Billingham (ed.), *Life in the universe*, pp. 297–306. MIT Press, Cambridge, Mass.

Tynni, R. (1980). Fossils in Lower Cambrian sedimentary sequences below thrust planes of Caledonian age in the Porojärvi area, Northern Finland. *Geologi*, **32**, 17–22. (In Finnish, with English Summary.)

Ulrich, W. (1972). Die Geschichte des Archicoelomatenbegriffs und die Archicoelomatennatur der Pogonophoren, *Z. Zool. Syst. Evol.-Forsch.*, **10**, 301–20.

Urbanek, A. & Mierzejewska, G. (1977). The fine structure of zooidal tubes in Sabelliditada and Pogonophora with reference to their affinity. *Acta Paleont. Polonica*, **22**, 223–39.

Valentine, J. W. (1973a). *Evolutionary paleoecology of the marine biosphere.* Prentice-Hall, Englewood Cliffs, N.J.

Valentine, J. W. (1973b). Coelomate superphyla. *Syst. Zool.*, **22**, 97–102.

Valentine, J. W. (1977). General patterns in metazoan evolution. In: A. Hallam (ed.), *Patterns of evolution*, pp. 27–57. Elsevier, Amsterdam.

Valentine, J. W. & Moores, E. M. (1972). Global tectonics of the marine biosphere. *J. Geol.*, **80**, 167–84.

Van Eysinga, F. W. B. (1975). *Geological time table* (3rd edn.) Elsevier, Amsterdam.
Veizer, J. (in press). The evolving terrestrial exogenic cycle. In: R. M. Garrels, C. B. Gregor & F. T. Mackenzie (eds.), *Chemical cycles in the evolution of the earth.* Wiley, New York.
Vidal, G. (1979). Acritarchs from the Upper Proterozoic and Lower Cambrian of East Greenland. *Grønlands Geol. Unders. Bull.*, **134**, 1–56.
Vidal, G. (1981). Aspects of problematic acid-resistant, organic-walled microfossils (acritarchs) in the Upper Proterozoic of the North Atlantic region. *Precambrian Res.*, **15**, 9–23.
Vogel, K. & Gutmann, W. F. (1981). Zur Entstehung von Metazoen-Skeletten an der Wende vom Präkambrium zum Kambrium. In: *Festschrift d. wissensch. Ges. J. W. Goethe Univ. Frankfurt a.M.*, pp. 517–37. F. Steiner Verlag, Wiesbaden.
Vologdin, A. G. & Maslov, A. B. (1960). (A new group of fossil organisms from the base of the Yudoma Series of the Siberian Platform.) *Doklady Akad. Nauk* SSSR, **134**, 691–3. (In Russian.)
Wade, M. (1968). Preservation of soft-bodied animals in Precambrian sandstones at Ediacara, South Australia. *Lethaia*, **1**, 238–67.
Wade, M. (1969). Medusae from uppermost Precambrian or Cambrian sandstones, central Australia. *Palaeontology*, **12**, 351–65.
Wade, M. (1970). The stratigraphic distribution of the Ediacara fauna in Australia. *Trans. Roy. Soc. S. Aust.*, **94**, 87–104.
Wade, M. (1971). Bilateral Precambrian chondrophores from the Ediacara fauna, South Australia. *Proc. Roy. Soc. Victoria*, **84**, 183–8.
Wade, M. (1972a). Hydrozoa and Scyphozoa and other medusoids from the Precambrian Ediacara fauna, South Australia. *Palaeontology*, **15**, 197–225.
Wade, M. (1972b). *Dickinsonia*: polychaete worms from the Late Precambrian Ediacara fauna, South Australia. *Mem. Queensland Mus.*, **16**, 171–90.
Wade, M. (in press). Fossil Scyphozoa. In: P. Grassé (ed.), *Traité de zoologie*. Masson et Cie, Paris.
Walcott, C. D. (1910). Cambrian geology and paleontology II. No. 1. Abrupt appearance of the Cambrian fauna of the North American Continent. *Smithsonian Inst. Misc. Coll.* **57**, 1–16.
Walker, J. C. G. (1977). *Evolution of the atmosphere.* Macmillan, New York.
Walter, M. R. (1972). Tectonically deformed sand volcanoes in a Precambrain greywacke, Northern Territory of Australia. *J. Geol. Soc. Aust.*, **18**, 395–9.
Walter, M. R. (1976). *Stromatolites.* (Developments in sedimentology, vol. 20.) Elsevier, Amsterdam.
Walter, M. R. (1977). Interpreting stromatolites. *Am. Scient.*, **65**, 563–71.
Walter, M. R. (1978). Recognition and significance of Archaean stromatolites. In: J. E. Glover & D. I. Groves (eds.), *Archaean cherty metasediments, their sedimentology, micropalaeontology, biogeochemistry, and significance to mineralization*, pp. 1–10. (Publs. Geol. Dept. & Ext. Serv. Univ. W. Aust., No. 2.) Nedlands, W. Australia.
Walter, M. R. (1983). Archaean stromatolites: evidence of the earth's earliest benthos. In: J. W. Schopf (ed.), *The earth's earliest biosphere: its origin and evolution*, Chapter 4. Princeton Univ. Press. Princeton, N.J.
Walter, M. R., Oehler, J. H. & Oehler, D. Z. (1976). Megascopic algae 1300 m.y. old from the Belt Supergroup, Montana: a reinterpretation of Walcott's *Helminthoidichnites*. *J. Paleont.* **50**, 872–81.
Wang Yuelun, Xing Yusheng, Lin Weixing, Zhang Luyi, Lu Zongbin, Gao Zhenjia, Ma Guogan & Lu Songnian, (1980). Sudivision and correlation of the Upper Precambrian in China. *Research on Precambrian Geology–Sinian Suberathem in China.* Tianjin Sci. Technol. Press. (In Chinese.) 407 pp.
Webby, B. D. (1970). Late Precambrian trace fossils from New South Wales. *Lethaia*, **3**, 79–109.

Webby, B. D. (1973). Trace fossils from the Lintiss Vale Formation of New South Wales: a Late Precambrian fauna. *Search*, **4**, 494–6.

Wegener, A. (1924). *The origin of continents and oceans.* Methuen, London. 212 pp.

Wegener, A. (1929). *The origin of continents and oceans.* Dover, New York. 236 pp. (In 1967 a translation of the 4th German edition of 1929 was published under the same title by Methuen, London, 248 pp.)

Werner, B. (1966). *Stephanoscyphus*, (Scyphozoa, Coronatae) und seine direkte Abstammung von den fossilen Conulata. *Helgoländer Wiss. Meeresunters.*, **13**, 317–47.

Werner, B. (1975). Bau und Lebensweise des Polypen von *Tripedalia cystophora* (Cubozoa class. nov., Carybdeidae) und seine Bedeutung für die Evolution der Cnidaria. *Helgoländer Wiss. Meeresunters.*, **27**, 561–504.

Werner, B. (1976). Die neue Cnidarierklasse Cubozoa. *Verh. Deutsch Zool. Ges.*, p. 230.

Weyl, P. K. (1968–69). Precambrian marine environment and the development of life. *Science*, **161**, 158–60, and **162**, 587 (reply).

Whittaker, R. H. (1969). New concepts of kingdoms of organisms. *Science*, **163**, 150–60.

Whittaker, R. H. & Margulis, L. (1978). Protist classification and the kingdoms of organisms. *BioSystems*, **10**, 3–18.

Whittington, H. B. (1979). Early arthropods, their appendages and relationships. In: M. R. House (ed.), *The origin of major invertebrate groups*, pp. 253–8. Academic Press, London.

Whittington, H. B. (1980). The significance of the fauna of the Burgess Shale, Middle Cambrian, British Columbia. *Proc. Geol. Ass.*, **91**, 127–48.

Williams, G. E. (1975). Late Precambrian glacial climate and the earth's obliquity. *Geol. Mag.*, **112**, 441–544.

Williams, H. & King, A. F. (1979). *Trepassey map area, Newfoundland, Geol. Survey Canada*, Mem. 389, 24 pp.

Windley, B. F. (1976). *The early history of the earth.* Wiley, New York.

Windley, B. F. (1977). *The evolving continents.* Wiley, London.

Wright, A. D. (1979). Brachiopod radiation. In: M. R. House (ed.), *The origin of major invertebrate groups*, pp. 235–52. Academic Press, London.

Yochelson, E. L. (1978). An alternative approach to the interpretation of the phylogeny of ancient molluscs. *Malacologia*, **17**, 165–91.

Yochelson, E. L. (1979). Charles D. Walcott – America's pioneer in Precambrian paleontology and stratigraphy. In: W. O. Kupsch & W. A. S. Sarjeant (eds.), *History of concepts in Precambrian geology*, Geol. Ass. Canada Spec. Paper 19, pp. 261–2.

Yochelson, E. D. & Stanley, G. D. (1981). An Early Ordovician patelliform gastropod, *Palaeolophacmaea*, reinterpreted as a coelenterate. *Lethaia*, **15**, 323–30.

Young, F. G. (1972). Early Cambrian and older trace fossils from the Southern Cordillera of Canada. *Canadian J. Earth Sci.*, **9**, 1–17.

Zhamoida, A. I. (ed.). (1979). *Stratigraphic code of the USSR. Provisional synopsis of rules and recommendations.* Ministry of Geology and Academy of Science, USSR, Leningrad.

Zhuravleva, I. T. (1970). Marine faunas and Lower Cambrian stratigraphy. *Am. J. Sci.*, **269**, 417–45.

Zhuravleva, I. T., Meshkova, N. P., Luchinina, V. A., Kashina, L. N., Korshunov, V. I. & Pel'man, Yu. L. (1979). (Peculiarities of the organic world at the Cambrian–Precambrian boundary (Northern Siberian Platform and neighbouring territories).) In: *Palaeontology of the Precambrian and Early Cambrian*, pp. 193–200. Inst. Geol. Geochron. Precambrian, Acad. Sci. USSR, 'Nauka', Leningrad. (In Russian.)

Ziegler, A. M., Scotese, C. R., McKerrow, W. S., Johnson, M. E. & Bambach, R. K. (1979). Paleozoic paleogeography. *Ann. Rev. Earth Planet. Sci.* **7**, 473–502.

Zonenshain, L. P. & Gorodnitsky, A. M. (1977). (Paleozoic and Mesozoic reconstructions of continents and oceans I. Early and mid-Palaeozoic reconstructions.) *Geotektonika*, No. 2, 3–23.

Zuckerkandl, E. (1965). The evolution of hemoglobin. *Sci. Am.*, **212**, 110–18.
Zuckerkandl, E. & Pauling, L. (1962). Molecular disease, evolution and genic heterogeneity. In: M. Kasha & B. Pullman, (eds.), *Horizons in biochemistry*, pp. 189–225. Academic Press, London.

## Dictionaries of scientific terms

### A. Biology

Abercrombie, M., Hickman, C. J. & Johnson, M. L. (1973). *A dictionary of biology* (6th edn). Penguin Books, Harmondsworth, England.

Gray, P. (1967). *The dictionary of the biological sciences.* Reinhold Book Co., New York.

Holmes, S. (ed.) (1979). *Henderson's dictionary of biological terms* (9th edn). Longmans, London.

Lapedes, D. N. (ed.) (1976). *McGraw Hill dictionary of life sciences.* McGraw Hill, New York.

### B. Earth Sciences

American Geological Institute (1976). *Dictionary of geological terms* (revised edn). Anchor Books, Garden City, New York.

Bates, R. L. & Jackson, J. A. (1980). *Glossary of geology.* Am. Geol. Inst., Falls Church, Virginia.

Challinor, J. (1973). *A dictionary of geology* (4th edn). Univ. Wales Press, Cardiff.

Stiegler, S. E. (ed.) (1976). *A dictionary of earth sciences.* Pan Book Ltd. (Macmillan, London.)

## Postscript

While this study was in the press, there appeared a book by W. B. Harland, A. V. Cox, P. G. Llewellyn, C. A. G. Pickton, A. G. Smith and R. Walters, *A geologic time scale*, Cambridge University Press, 1982. As there is in my Sections 1.2 and 4.1 much discussion of geologic time scales which are essential for the study of the history of life and the earth, this new book will be of much interest to my readers. My comments on its treatment of the relevant time interval, the Precambrian–Cambrian transition, could not be incorporated in this text. They will appear in the *Geological Magazine* in 1984. For this, my sincere thanks are due to John Cowie and Brian Harland.

Adelaide, May 1983

# Author index

Note: Only the first-named of multiple authors mentioned in the text are included here. The alphabetic listings of names under the heading 'References' above are not repeated.

Abel, O., 198
Alport, S. P., 182, 184
Anderson, M. M. 93, 98
Asklund, B., 216
Avnimelech, M., 126
Awramik, S. M., 20, 178
Ayala, F. J., 199

Banks, J., 73
Banks, N. L., 183
Barghoorn, E. S., 4
Barnes, R. D., 134
Beckinsale, R. D., 92
Beer, E. J., 26
de Beer, G. N., 32
Bekker, Yu. R., 87
Beklemishev, V. N., 33, 134
Bengtson, S., 158, 159, 160, 163, 165
Berkner, L. V., 7, 8, 169
Bernal, J. D., 2, 3
Beuf, S., 206
Beurlen, K., 81
Binda, P. L., 20
Birket-Smith, S. J. R., 61, 122
Bischoff, G. C. O., 54, 113, 114, 115
Bloeser, B., 20
Boaden, P. J. S., 38, 39
Bock, W. J., 198
Boltzmann, L., 17
Borello, A., 83
Borovikov, L. I., 144
Bouillon, J., 113
Boyden, A., 32
Brasier, M. D., 26, 93, 155, 156, 158, 176, 177, 178, 180, 187
Brien, P., 31
Broda, E., 3, 5, 10, 15, 17, 18

Calvin, M., 4, 10
Cavalier-Smith, T., 15
Chang, S., 2
Chumakov, N. M., 203, 204
Chung, F. T., 90
Cisne, J. L., 120
Clark, R. B., 36, 40, 41, 42, 62, 110, 118, 119, 121, 123, 125, 126, 127, 136, 173, 178, 187, 188
Clemmey, H., 25
Cloud, P., 7, 8, 9, 12, 15, 16, 18, 25, 26, 27, 69, 70, 98, 99, 100, 104, 140, 143, 169, 181, 200, 201, 202, 203, 214
Coats, R. P., 44, 47
Cohee, G. V., 2
Conway Morris, S., 61, 121, 122, 126, 127, 153, 156, 162, 164, 179
Cope, J. C. W., 93
Cowie, J. W., 151
Crawford, A. R., 203, 205
Cribb, S. J., 92, 93
Crimes, T. P., 182, 185, 186, 187

Daily, B., 44, 47, 70, 150, 153, 155, 182
Dalgarno, C. R., 46
Dalla Salda, L., 81
Darwin, C., 10, 109
Debrenne, F., 160, 161
Degens, E. T., 170
Derstler, K., 65
Ding, Q. X., 90
Donaldson, J. A., 22
Dougherty, E., 31
Downie, C., 25
Dunning, F. W., 92, 93
Durham, J. W., 26, 158, 159, 160
Du Toit, A. L., 208

# Author index

Eldredge, N., 31, 194
Evitt, W. R., 25

Fairbridge, R. W., 206
Fairchild, T. R., 20, 81
Fauchald, K., 63
Fedonkin, M. A., 56, 61, 84, 87, 88, 122, 132, 149, 168, 183, 188, 199, 202
Fenchel, T. M., 37, 127, 178
Fenton, M. A., 126
Fischer, A. G., 7
Florkin, M., 10, 167, 168
Flounders, B., 43, 44
Forbes, B. G., 44, 47
Ford, T. D., 25, 56, 92, 93, 95
Foster, M. W., 156
Fox, S. W., 4
Føyn, S., 73, 155, 167, 183
Frakes, L. A., 203
Francis, S., 19
Frey, R. W., 188
Fritz, W. H., 153, 179
Fyodorov, A. B., 165

Garrett, P., 178
Gebelein, C. D., 23
Germs, G. J. B., 57, 59, 73, 74, 76, 77, 81, 161, 167, 185
Glaessner, M. F., 2, 12, 25, 27, 30, 44, 47, 54, 55, 56, 57, 64, 65, 69, 71, 77, 80, 110, 111, 113, 115, 119, 136, 148, 159, 161, 167, 181, 185
Goethe, J. W., 198
Goldring, R., 49
Gould, S. J., 31, 32, 190, 195, 196
Grasshoff, M., 122, 124
Grell, K. G., 18
Gürich, G., 56, 73, 81
Gutmann, W. F., 37, 126, 129, 173, 198

Hadži, J., 36
Haeckel, E., 32, 33, 35
Hahn, G., 99
Hakes, W. B., 66
Haldane, J. B. S., 2, 3
Hallam, A., 31, 195
Hambrey, M. J., 98, 203
Hand, C., 36
Hanson, E. D., 19, 20, 31, 32, 33, 35, 36, 110, 128, 196
Häntzschel, W., 65
Harland, W. B., 139, 141, 143, 203, 205
Harrison, J. E., 139
Haughton, S. H., 77
Heberer, G., 31
Hedberg, H. D., 2, 6, 139
Henderson, L. J., 214
Hofmann, H. J., 11, 25, 96, 99, 153, 181

Holland, H. D., 200, 208, 215
House, M. R., 110
Houzay, J.-P., 156
Howard, J. D., 184
Hutchinson, G. E., 15
Hyman, L. H., 35, 36, 39, 48, 162

Ivanov, A. V., 33, 34, 35, 36, 39, 112

Jaeger, H., 185
Jägersten, G., 31, 36, 39, 119, 196
James, H. L., 6, 140
Jell, P. A., 163, 164
Jenkins, R. J. F., 45, 46, 47, 49, 56, 69, 70, 112, 157
Jeuniaux, Ch., 167

Kauffman, E. G., 26
Keller, B. M., 83, 142
Kent, L. E., 74
Knoll, A. H., 19, 20
Kovalevsky, A., 33
Kowalski, W. R., 185
Kretsinger, R. H., 170, 171
Kröner, A., 74, 203, 204, 205, 206
Kulling, O., 90

Lankester, E. R., 35
Larwood, G. P., 126
Lauterbach, K. E., 122, 123, 124
Leeson, B., 44
Liu, X. L., 90
Lovelock, J. E., 215
Lowenstam, H. A., 168, 170, 171

McKirdy, D. M., 11
McLaughlin, P. J., 10
McWilliams, M. O., 205
Major, R. B., 70
Manton, S. M., 122, 129
Marek, L., 159, 163
Margulis, L., 7, 15, 18, 40, 200, 201
Martin, H., 74
Matthews, S. C., 158
Mawson, D., 43, 44
Mayr, E., 197, 199
Meneisy, M. Y., 92
Metchnikoff, E., 33, 35, 36
Mettam, L., 67
Miller, S. L., 2
Mincham, H., 43
Misra, S. B., 95, 97
Missarzhevsky, V. V., 151, 152, 153, 158, 162, 163
Monod, J., 198
Morel, P., 131, 205
Müller, A. H., 13
Müller, K. J., 110, 122

Nitecki, M. G., 160
Nursall, J. R., 107

Oehler, J. H., 19
Oparin, A. I., 2, 3
Öpik, A. A., 160
Osgood, R. G., 184

Palij, V. M., 26, 87
Pantin, F. C. A., 32
Patchett, P. J., 92
Patterson, C., 7
Paul, C. R., 65
Peters, D. S., 32
Pflug, H. D., 4, 56, 57, 73, 77, 80, 81, 83
Pojeta, J., 163, 173
Poulsen, V., 160
Preiss, W. V., 22
Pringle, I. R., 143

Raff, R. A., 186
Range, P., 73
Ratner, M. I., 18, 201
Reisinger, E., 37
Remane, A., 32, 39, 110
Repetski, J. E., 163
Rhoads, D. C., 37, 38, 127, 178, 186
Richter, R., 56, 73, 81
Riedl, L., 31, 32, 196, 197, 198
Rieger, R. M., 125
Rieppel, O., 32
Roberts, J. D., 203
Rowell, A. J., 126
Rozanov, A. Yu., 131, 150, 151, 158, 159, 160, 165
Runnegar, B., 60, 163, 165, 173
Rutten, M. G., 7, 170

Sabrodin, W. (Zabrodin), 24, 25
Salvini-Plawen, L. v., 42, 112, 113, 115, 125
Scheltema, A. H., 125
Schneiderhöhn, H., 73
Schopf, J. W., 4, 14, 16, 17, 29, 202
Schopf, T. J. M., 195, 196, 198, 201, 206, 207, 208
Schrödinger, E., 17
Scrutton, C. T., 113, 114
Seilacher, A., 67, 187
Semikhatov, M. A., 22
Sepkoski, J. J., 147, 148, 149, 151, 158, 176, 187, 188, 189, 190, 191, 192, 193, 201, 202, 203, 206, 207
Sewertzoff, A. N., 197
Siewing, R., 37
Simpson, G. G., 32, 119, 132, 133, 134, 136, 196

Sims, P. K., 6, 140
Sokolov, B. S., 25, 83, 87, 88, 99, 121, 140, 142, 143, 144, 146, 148, 149, 153, 168
Sprigg, R. C., 43, 44, 47, 48, 56, 112
Squire, A. D., 25, 121
Stanier, R., 15
Stanley, S. M., 30, 31, 109, 135, 136, 175, 178, 191, 194, 202, 207
Stasek, C. R., 163, 164
Stöcklin, J., 99
Szaniawski, H., 163

Tankard, A. J., 74–6
Tappan, H. N., 14, 15, 24
Tarling, D. H., 6
Taylor, F. J. R., 15
Termier, H., 100, 143
Thiel, M. E., 70
Timofeev, B. V., 25
Towe, K. M., 131, 172, 176
Tynni, R., 92, 155

Ulrich, W. 37, 120
Urbanek, A., 121

Valentine, J. W., 118, 134, 202, 209, 210, 211
Van Eysinga, F. W. B., 2
Veizer, J., 170, 208
Vidal, G., 20, 25
Vogel, K., 124, 173, 174, 186

Waddington, C. H., 32, 197
Wade, M., 46, 50, 53, 54, 55, 56, 60, 61, 112, 114, 115, 117
Walcott, C. D., 149, 150
Walter, M. R., 4, 22, 26, 70, 73
Wang, Y., 90, 121
Webby, B. D., 182
Wegener, A., 208, 209
Werner, B., 54, 114, 115
Weyl, P. K., 179
Whittaker, R. H., 20, 21, 22, 33
Whittington, H. B., 110, 123, 156, 179, 190
Williams, G. E., 203, 204, 205, 206
Williams, H., 93, 95, 98
Windley, B. F., 6, 209
Wright, A. D., 126, 162

Yochelson, E. L., 146, 150, 165
Young, F. G., 130

Zhamoida, A. I., 216
Zhuravleva, I. T., 158, 159, 162, 181
Ziegler, A. M., 131, 205, 209
Zonenshain, L. P., 131, 205
Zuckerkandl, E., 10

# Subject index

Note: Subjects included in chapter and section headings are generally omitted, as are most geographic and stratigraphic terms. Most of them can be found in the list of figures and tables at the beginning of the book. The inclusion of zoological terms, morphological and systematic, except names of genera, would have greatly increased the length of this index without adding significantly to its usefulness.

Acritarcha, 25
Actualism, 208, 211
Adenosine triphosphate (ATP), 17
Age dating (geochronology) of
    Archaean–Proterozoic boundary, 6
    Charnian (English Midlands), 92
    Conception Group (Newfoundland), 98, 104
    Ediacarian, 104, 145
    Late Precambrian, 102–4, 142, 143, 204
    Mid-Precambrian, 7
    Nama Group (southwestern Africa), 74
    Vendian (USSR), 88, 143
*Albumares*, 84, 86
*Aldanella*, 155
Algae, 4, 16, 88, 128, 153, 200
    blue-green (*see also* Cyanobacteria. Cyanophyta), 4, 13, 15, 16, 18, 19, 22, 179, 217
    green, 160
    planktonic, 24
*Anabarella*, 153
*Anabarites*, 90, 152, 153, 155, 158, 160
*Ancalogon*, 127
Animal protists, *see* Protozoa
*Archaeichnium*, 77, 167, 185
*Arenicolites*, 93
*Arumberia*, 73, 87, 90
*Asterosoma*, 26
*Astrapolithon*, 93
Atmosphere
    carbon dioxide content, 169
    composition, 200
    oxygen content, *see* Oxygen
    reducing, 3

*Aulophycus*, 81
*Aysheaia*, 123

Bacteria, 4, 5, 13, 15, 16, 18, 22, 146, 200
*Baikalina*, 88, 90
Banded iron formations (BIF), 8, 16
*Bathysiphon*, 167
*Beltanelloides*, 24, 25
*Bemella*, 153
*Bergaueria*, 185, 187, 188
'Bilaterogastraea' theory, 35, 39
*Bilinichnus*, 183
Biomass, 156, 178, 191, 202
Biomechanics, 172, 173, 185, 197
Biostratigraphy, 5, 208, 209
Bioturbation, 25, 26, 121, 127, 200
Bitter Springs microbiota, 19, 20
*Brachina*, 54, 70, 88
*Brooksella canyonensis*, 26
*Buchholzbrunnichnus*, 185
*Bunyerichnus*, 69

*Calyptrina*, 89
'Cambrian fauna' (*sensu* Sepkoski), 188, 190
*Cambridium*, 14
*Cambrotubulus*, 152, 153, 158, 160
*Camenella*, 155
Cell, 3, 4, 13, 33
    division, 7, 13
    membrane, 3, 19
    size, 13, 14
    walls, 4, 19
*Charnia*, 57, 84, 87, 90, 91, 92, 93, 95, 99, 100, 155, 183

*Charniodiscus*, 56, 57, 69, 70, 71, 84, 87, 92, 93, 95, 127, 168, 183
Chemical evolution, 4, 10
'Chemical fossils', 11
Chert, 11, 19
Chitin, 4, 19, 133, 167, 168, 197
Chitinozoa, 20, 25
Choanoflagellates, 33, 34, 35, 40
Chronology (*see also* Age dating), 5
*Chuaria*, 25, 121
Ciliates, 20
Classification, principles of, 132, 133
*Climactichnites*, 185
Climate, 205, 206, 210
*Cloudina*, 74, 76, 77, 78, 81, 167
Coacervates, 4
Coccolithophorids, 24
*Cochlichnus*, 182, 183
Coelomate radiation, *see* Evolutionary radiation
*Coleoloides*, 155
Collagen, 131, 172, 176
Community structure, 180, 199, 200
*Conchopeltis*, 54, 146
Configurations, inorganic, mechanical, resembling fossils, 12
*Conomedusites*, 54, 115, 146, 168
*Cruziana*, 184, 185, 187
*Curvolithus*, 182, 184
Cyanobacteria, (*see also* Algae, blue-green; Cyanophyta), 13, 217
Cyanophyta (*see also* Algae, blue-green; Cyanobacteria), 5, 19, 25, 217
*Cyclomedusa*, 54, 67, 74, 88, 90, 92
*Cylindrichnus*, 66

Diatomacea (diatoms), 25, 208
*Dickinsonia*, 57, 60, 61, 84, 86, 87, 119, 127, 129, 134, 136, 137, 144
*Didymaulichnus*, 129, 130, 184, 185
*Dimorphichnus*, 182
Dinoflagellata, 25
*Diplichnites*, 184, 187
*Diplocraterion*, 46, 181, 187
Diversification, rates of, 135, 192
Diversity (faunal, taxonomic), 131, 132, 148, 149, 176, 196
DNA (Deoxyribonucleic acid), 4, 14
Dysaerobic, 39

*Ediacaria*, 54
Energy
 conservation, 174
 for prebiotic synthesis, 2, 3
 in the form of light, 17
 utilization, 196
*Ernietta*, 77, 80, 83
Eukaryota, 8, 14, 17
 origins, first appearance, 7, 15, 16, 18, 19, 26, 29, 201, 202

Evolution
 biochemical, 17, 168
 biospheric, 9, 208
 cytological, 29, 30
 Darwinian, 135
 factors determining, 191, 196, 197, 198
 geochemical, 8, 208
 gradualistic model, 136
 opportunistic, 32, 119, 123, 133, 135, 192, 203
 punctuational model, 135, 136, 194
 rates, 17, 135, 137, 184, 187, 192, 193, 194, 203
Evolutionary radiation
 acoelomate, 118, 202, 203
 arthropod, 121, 123
 cnidarian, 112
 coelomate, 41, 118, 119, 121, 122, 123, 127, 172, 203
 deuterostome, 118
 lophophorate, 118, 125
 pre-coelomate, 202

Fecal
 castings, 184
 matter, 200
 pellets, 24, 25, 26, 121, 129, 202
Fermentation, 16, 196, 200
*Fomitchella*, 159, 163
Foraminifera, 12, 20, 128, 166, 167
Form–function complex, 198
Fossilization, 11, 13, 50, 57, 98, 133, 167
Fossil record, incompleteness of, 30, 108ff, 171
Frond-like fossils, 56, 93, 112
Functional morphology, 32, 36, 185

Genetic code, 4, 213
Geochronology, *see* Age dating
'Geochronometric scale', 139
'Geochronostratic scale', 139
*Girvanella*, 153
Glaciations (Late Precambrian), 29, 30, 102–4, 108, 175, 203, 204, 205, 213
*Glaessnerina*, 57, 69, 87, 88, 90, 91, 149, 153
*Gordia*, 93
Gunflint microbiota, 19, 20, 33

*Halkieria*, 164
*Hallidaya*, 54, 73, 84
*Harlaniella*, 183
*Helcionella*, 153
*Helminthoidichnites*, 184
Holozoic (Eon), 141, 142
Homeostatic, 32, 197, 215
Homology, 32, 110, 111
Hydrostatic skeleton, 127, 172
*Hyolithellus*, 155, 159, 160
*Hyolithes*, 155

## Subject index

Information theory, 198
*Inkrylovia*, 87, 99

Kerogen, 11
*Kimberella*, 54, 114, 117
*Kullingia*, 73, 90, 92

*Laetmonice*, 63, 64
*Lenargyrion*, 166
*Lorenzinites*, 55

*Majella*, 88
*Margaritichnus*, 65
*Marywadea*, 61
*Mawsonites*, 55
*Medusinites*, 55
Metazoa
  distribution in Late Precambrian, 27
  diversification, 27, 107, 197, 201, 203, 205, 212
  emergence, 196, 198, 200
  evolution, 112, 119, 137
  origin, 7, 18, 19, 20, 29, 33, 36, 112
Meteorites, 2, 6
*Micromitra*, 155
Microplankton, 37, 63, 179
Microspheres, 4
Mineralized tissues, 27, 30, 146, 168, 172, 179
*Mobergella*, 94, 155, 156, 166
'Molecular palaeontology', 10
Molecules, organic, 2

*Namalia*, 77, 88, 113
*Nasepia*, 77
*Nemiana*, 85
*Nenoxites*, 183
*Neonereites*, 66, 183
Neoteny (paedomorphism), 118, 119
*Nimbia*, 84

Obliquity of the ecliptic, 204, 205, 206
*Obolella*, 94, 155
*Odontogriphus*, 163
Oxygen
  atmospheric, 7, 9, 16, 131, 169, 170, 172, 173, 175, 179, 183, 196, 197, 200, 201, 208
  present atmospheric level, *see* PAL
  requirements, 172-3, 176, 200, 201
  toxicity, 18
Ozone, 17, 131, 201

PAL (present atmospheric level of oxygen), 8, 9, 16, 18, 39, 169, 175, 186, 197, 200, 201
Palaeobiology, 30, 77, 118, 198
Palaeoclimatology, 211
Palaeoecology, 27, 101, 130–2, 151, 192, 207

Palaeogeography, 131, 175, 180, 205, 109, 211
Palaeo-oceanography, 179, 180, 211
*Palaeopascichnus*, 183
'Palaeozoic fuana' (*sensu* Sepkoski), 188, 190, 191, 206
*Paleolina*, 89, 160
*Paramedusium*, 77
Parenchymella, *see* Phagocytella
*Parvancorina*, 64, 85, 121, 123, 168
'Pasteur point', 7, 186, 201
*Persimedusites*, 99
Petalonamae, 56, 77, 112, 148
Phagocytella (= Parenchymella), 35
Phosphate deposits, 200
Phosphate mineralization, 170
Photosynthesis, 13, 16, 21, 200, 215
*Phycodes*, 74, 182
*Phyllozoon*, 57
Phylogeny, 31, 33, 36, 40, 109, 133
Phytoplankton, 24, 39
*Plagiogmus*, 182, 185
*Planolites*, 70, 81, 93, 182, 183, 185
Plant flagellates
  (= phytoflagellates = Phytomonadina), 18, 21
Planula, planuloid, 35, 36, 112
*Platysolenites*, 90, 155, 167
*Plumulites*, 164
Porifera, origin of, 33, 35, 40, 128, 160, 166
*Porpita*, 51, 53
*Praecambridium*, 64, 111, 129, 168
Predators, 29, 37, 67, 128, 146, 174, 191, 199, 200, 207
Prehistory, metazoan, 30, 106, 194, 202, 203, 212
Prokaryota, 4, 7, 8, 13, 14, 15, 33, 37, 212
*Protechiurus*, 77, 83, 134
*Proterospongia*, 35
Protista, 18, 20, 21, 29, 33, 160, 196, 203, 212, 214
*Protodipleurosoma*, 56
*Protohertzina*, 120, 146, 152, 153, 155, 158, 160, 162, 163
Protozoa, 7, 18, 19, 20, 33, 128, 214
*Psammichnites*, 93
Pseudofossils, 11, 12, 69, 90
*Pseudorhizostomites*, 55, 56, 85
*Pteridinium*, 57, 77, 79, 80, 84, 85, 86, 88, 183

Quantum evolution, 136, 196

Radioastronomy, 2
Radiolaria, 12, 20, 128, 170, 208
*Rangea*, 57, 77, 81, 95
*Redkinia*, 87, 88, 168
Respiration, 16, 17, 29, 172, 173, 174, 175, 196, 200, 201

*Rugoconites*, 55, 100
*Rugoinfractus*, 26
*Rusophycus*, 90, 165, 174, 184

*Sabellidites*, 90, 152, 153, 155, 158, 159, 160
*Sachites*, 164
Sclerites, 160, 162, 163, 164, 166, 173
'Scolicia', 184
Shells, skeletons, 13, 109, 125, 128, 135, 150, 151, 159, 162, 163, 168, 172, 174, 176, 180, 197
Silicification, 80, 81
Skeletons, *see* shells
*Skinnera*, 73, 83
*Skolithos*, 67, 73, 81, 120, 126, 181, 184, 187, 188
Soft-bodied organisms, 13, 102, 132, 146, 156, 157, 162, 172, 190
  fossilization potential of, 130
Speciation, 135, 136
Spicules, 125, 159, 168, 169
  sponge, 40
*Spinther*, 57, 61
*Spriggina*, 61, 62, 63, 119, 122, 129
Stephanoscyphus, 54, 114
Stromatolites, 4, 17, 22, 23, 37, 142, 178
Structural plan (= body plan, groundplan, morphotype), 32, 119, 123, 135, 174, 192, 213
Sulfide biome (= thiobiome), 36, 38, 127, 128
*Sunnaginia*, 155
Supercontinent, 209, 210
*Suvorovella*, 88
*Suzmites*, 70, 183
Symbiotic theory of eukaryote origin, 15
System theory, 198

*Tannuolina*, 164
*Tawuia*, 121
*Teichichnus*, 185
*Thambetolepis*, 164
Thermocline, 179, 180
Thigmotactic grazing, 66, 129
Thiobios, *see* Sulfide biome
Tillite, Late Precambrian, 98, 102, 142, 143, 206
*Tirasiana*, 74, 85, 87
*Tomopteris*, 122
*Torellella*, 160
*Torrowangea*, 182, 183, 185
Trace fossils, 12, 29, 65, 74, 76, 81, 83, 87, 88, 90, 93, 110, 121, 123, 125, 126, 129, 130, 133, 148, 158, 165, 180–8
Transgression and regression, 150, 155, 175, 176, 177, 178, 210
*Treptichnus*, 185
*Tribrachidium*, 65, 85, 87, 120, 125, 134, 146, 148
*Tripedalia*, 114

Ultraviolet radiation, 2, 17, 18, 53, 131, 201

*Velella*, 51, 53
*Velumbrella*, 93
*Vendia*, 85, 87, 122
*Vendichnus*, 183
*Vendomia*, 85, 87
*Vermiforma*, 98
Virus, 3, 13
*Volborthella*, 155

*Wiwaxia*, 164

Zooflagellates, 20, 33